SpringerBriefs in Applied Sciences and Technology

PoliMI SpringerBriefs

More information about this subseries at https://link.springer.com/bookseries/11159
http://www.polimi.it

Luigi Piroddi
Editor

Special Topics in Information Technology

POLITECNICO
MILANO 1863

Editor
Luigi Piroddi
Dipartimento di Elettronica, Informazione e
Bioingegneria
Politecnico di Milano
Milano, Italy

ISSN 2191-530X   ISSN 2191-5318   (electronic)
SpringerBriefs in Applied Sciences and Technology
ISSN 2282-2577   ISSN 2282-2585   (electronic)
PoliMI SpringerBriefs
ISBN 978-3-030-85917-6   ISBN 978-3-030-85918-3   (eBook)
https://doi.org/10.1007/978-3-030-85918-3

This Springer imprint is published by the registered company Springer Nature Switzerland AG
The registered company address is: Gewerbestrasse 11, 6330 Cham, Switzerland

Preface

This volume collects some of the most promising results achieved by recently graduated PhDs in Information Technology (IT) at the Department of Electronics, Information and Bioengineering of the Politecnico di Milano. The presented contributions summarize the main achievements of their theses, successfully defended in the a.y. 2020–21 and selected for the IT Ph.D. Award (out of more than 50 theses defended this year). As in the tradition of this Ph.D. program, these theses cover a wide range of topics in IT, reflecting the broad articulation of the program in the areas of computer science and engineering, electronics, systems and control, and telecommunications, and emphasizing the interdisciplinary nature of IT. Doctoral studies in the IT Ph.D. program pursue state-of-the-art research and the development of innovative, cutting-edge methodologies and technologies, and aim at preparing generations of young researchers that will shape future innovation, as the 12 authors of this volume will undoubtedly do. Overall, this book gives an overview of some of the newest research trends in IT developed at the Politecnico di Milano, presenting them in an easy-to-read format, amenable also to non-specialists.

Milano, Italy   Luigi Piroddi
July 2021

Contents

Systems and Control

Deep Learning in Multi-step Forecasting of Chaotic Dynamics   3
Matteo Sangiorgio

Optimal Management and Control of Smart Thermal-Energy Grids . . .   15
Stefano Spinelli

Electronics

**Application Specific Integrated Circuits for High Resolution X
and Gamma Ray Semiconductor Detectors**   31
Filippo Mele

**Modeling of GIDL–Assisted Erase in 3–D NAND Flash Memory
Arrays and Its Employment in NOR Flash–Based Spiking Neural
Networks**   43
Gerardo Malavena

**Low-Noise Mixed-Signal Electronics for Closed-Loop Control
of Complex Photonic Circuits**   55
Francesco Zanetto

Computer Science and Engineering

**Beyond the Traditional Analyses and Resource Management
in Real-Time Systems**   67
Federico Reghenzani

**Computational Inference of DNA Folding Principles: From Data
Management to Machine Learning**   79
Luca Nanni

**Model, Integrate, Search... Repeat: A Sound Approach to Building
Integrated Repositories of Genomic Data** 89
Anna Bernasconi

**Configurable Environments in Reinforcement Learning:
An Overview** 101
Alberto Maria Metelli

Machine Learning for Scientific Data Analysis 115
Gabriele Scalia

Telecommunications

**Sensor-Assisted Cooperative Localization and Communication
in Multi-agent Networks** 129
Mattia Brambilla

**Design and Control Recipes for Complex Photonic Integrated
Circuits** 141
Maziyar Milanizadeh

Systems and Control

Deep Learning in Multi-step Forecasting of Chaotic Dynamics

Matteo Sangiorgio

Abstract The prediction of chaotic dynamical systems' future evolution is widely debated and represents a hot topic in the context of nonlinear time series analysis. Recent advances in the field proved that machine learning techniques, and in particular artificial neural networks, are well suited to deal with this problem. The current state-of-the-art primarily focuses on noise-free time series, an ideal situation that never occurs in real-world applications. This chapter provides a comprehensive analysis that aims at bridging the gap between the deterministic dynamics generated by archetypal chaotic systems, and the real-world time series. We also deeply explore the importance of different typologies of noise, namely observation and structural noise. Artificial intelligence techniques turned out to provide robust predictions, and potentially represent an effective and flexible alternative to the traditional physically-based approach for real-world applications. Besides the accuracy of the forecasting, the domain-adaptation analysis attested the high generalization capability of the neural predictors across a relatively heterogeneous spatial domain.

1 Introduction

Machine learning techniques are nowadays widely used in time series analysis and forecasting, especially in those characterized by complex nonlinear behaviors. A classical example are the meteorological processes, whose nonlinearity often generates chaotic trajectories. In such context, the machine learning algorithms proved to outperform the traditional methodologies, mainly relying on linear modelling techniques [1, 2].

Artificial neural networks are by far the most widespread technique used for time series prediction. Two different neural architectures can be adopted in this context. The first are the feed-forward (FF) neural networks, so-called because their structures are described by an acyclic graph (i.e., without self-loops). FF architectures are thus

M. Sangiorgio (✉)
Dipartimento di Elettronica, Informazione e Bioingegneria, Politecnico di Milano, Via Ponzio 34/5, 20133, Milano, Italy
e-mail: matteo.sangiorgio@polimi.it

© The Author(s) 2022
L. Piroddi (ed.), *Special Topics in Information Technology*,
PoliMI SpringerBriefs, https://doi.org/10.1007/978-3-030-85918-3_1

static and different strategies can be used to adapt them to the intrinsically dynamical nature of time series. A second alternative is represented by recurrent neural networks (RNNs), so-called because of the presence of self-loops that make them dynamical models. This feature should make, at least in principle, the RNNs naturally suited for sequential tasks as time series forecasting.

In this chapter, we investigate the predictive accuracy of different purely autoregressive neural approaches with both FF and recurrent structures. Our analysis takes into account various chaotic systems, spanning from the archetypal examples of deterministic chaos to two real-world time series of ozone concentration and solar irradiance. To quantify the effect of different noise typologies on the forecasting capabilities of the neural predictors, we set up two numerical experiments by perturbing the deterministic dynamics of the archetypal chaotic systems with artificially-generated noise.

Finally, we extend the idea of domain adaptation to time series forecasting: the neural predictors trained to forecast the solar irradiance on a given location (the source domain) are then used, without retraining, as predictors for the same variable in other locations (the target domains).

The rest of this chapter is organized as follows. Section 2 introduces the feed-forward and recurrent neural predictors, describing their structures and their identification procedures. Section 3 reports the results obtained in a deterministic environment. In Sect. 4, the impact of different noise typologies on the forecasting accuracy is evaluated. Section 5 presents two real-world applications. Finally, in Sect. 6, some concluding remarks are drawn.

2 Neural Predictors for Time Series

2.1 Predictors' Identification

The forecasting of a time series is a typical example of supervised learning task: the training is performed by optimizing a suitable metric (the loss function), that assesses the distance between N target samples, $\mathbf{y} = [y(1), y(2), \ldots, y(N)]$, and the corresponding predictions, forecasted i steps ahead, $\hat{\mathbf{y}}^{(i)} = [\hat{y}(1)^{(i)}, \hat{y}(2)^{(i)}, \ldots, \hat{y}(N)^{(i)}]$. A widely used loss function is the mean squared error (MSE), that can be computed for the ith step ahead:

$$\text{MSE}(\mathbf{y}, \hat{\mathbf{y}}^{(i)}) = \frac{1}{N} \sum_{t=1}^{N} \left(y(t) - \hat{y}(t)^{(i)} \right)^2. \tag{1}$$

One-step predictors are optimized by minimizing $\text{MSE}(\mathbf{y}, \hat{\mathbf{y}}^{(1)})$. Conversely, a multi-step predictors can be directly trained on the loss function computed on the entire h-step horizon:

$$\frac{1}{h} \sum_{i=1}^{h} \text{MSE}(\mathbf{y}, \hat{\mathbf{y}}^{(i)}). \qquad (2)$$

Following the classical neural nets learning framework, the training is performed for each combination of the hyper-parameters (mini-batch size, learning rate and decay factor), selecting the best performing on the validation dataset. The hyper-parameters specifying the neural structure (i.e., the number of hidden layers and of neurons per layer) are fixed: 3 hidden layers of 10 neurons each. Once the predictor has been selected, it is evaluated on the test dataset (never used in the previous phases) in order to have a fair performance assessment and to avoid overfitting on both the training and the validation datasets.

A well-known issue with the MSE is that the value it assumes does not give any general insight about the goodness of the forecasting. To overcome this inconvenient, the R^2-score is usually adopted in the test phase:

$$R^2\left(\mathbf{y}, \hat{\mathbf{y}}^{(i)}\right) = 1 - \frac{\text{MSE}(\mathbf{y}, \hat{\mathbf{y}}^{(i)})}{\text{MSE}(\mathbf{y}, \bar{y})}. \qquad (3)$$

Note that $\bar{y}$ is the average of the data, and thus the denominator corresponds to the variance, $\text{var}(\mathbf{y})$. For this reason, the R^2-score can be seen as a normalized version of the MSE. It is equal to 1 in the case of a perfect forecasting, while a value equal to 0 indicates that the performance is equivalent to that of a trivial model always predicting the mean value of the data. An R^2-score of -1 reveals that the target and predicted sequences are two trajectories with the same statistical properties (they move within the same chaotic attractor) but not correlated [3, 4]. In other words, the predictor would be able to reproduce the actual attractor, but the timing of the forecasting is completely wrong (in this case, we would say that we can reproduce the long-term behavior or the climate of the attractor [5, 6]).

2.2 Feed-Forward Neural Networks

The easiest approach to adapt a static FF architecture to time series forecasting consists in identifying a one-step predictor, and to use it recursively (FF-recursive predictor). Its learning problem requires to minimize the MSE in Eq. (1) with $i = 1$ (Fig. 1, left), meaning that only observed values are used in input.

Once the one-step predictor has been trained, it can be used in inference mode to forecast a multi-step horizon (of h steps) by feeding the predicted output as input for the following step (Fig. 1, right).

Alternative approaches based on FF networks are the so-called FF-multi-output and FF-multi-model. In the multi-output approach, the network has h neurons in the output layer, each one performing the forecasting at a certain time step of the horizon. The multi-model approach requires to identify h predictors, each one specifically

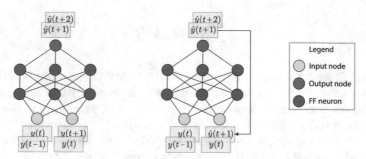

Fig. 1 FF-recursive predictor in training (**left**) and inference (**right**) modes. For the sake of simplicity, we considered $d = 2$ lags and $h = 2$ leads, and a neural architecture with 2 hidden layers of 3 neurons each

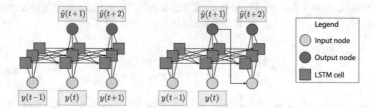

Fig. 2 RNNs trained with (**left**) and without (**right**) teacher forcing. Note that both LSTM-TF and LSTM-no-TF follow the scheme in panel b in inference mode, because at time t, the actual value at $t + 1$ in not available. For the sake of simplicity, we considered $d = 2$ lags and $h = 2$ leads, and a neural architecture with a single hidden layers of 3 cells

trained for a given time step. In this chapter, we limit the analysis to the FF-recursive predictor. A broad exploration including the other FF-based approaches can be found in the literature [7, 8].

2.3 Recurrent Neural Networks

Recurrent architectures are naturally suited for sequential tasks as time series forecasting and allow to explicitly take into account temporal dynamics. In this work, we make use of RNNs formed by long short-term memory (LSTM) cells, which have been successfully employed in a wide range of sequential task, from natural language processing to policy identification.

When dealing with multi-step forecasting, the RNNs are usually trained with a technique known as teacher forcing (TF). It requires to always feed the actual values, instead of the predictions computed at the previous steps, as input (Fig. 2, left).

TF proved to be necessary in almost all the natural language processing tasks, since it guides the optimization ensuring the convergence to a suitable parameterization. For this reason, it became the standard technique implemented in all the deep learning libraries. TF simplifies the optimization at the cost of making training and inference modes different. In practice, at time t, we cannot use the future values $y(t + 1)$, $y(t + 2)$, ... because such values are unknown and must be replaced with their predictions $\hat{y}(t + 1)$, $\hat{y}(t + 2)$, ... (Fig. 2, right).

This discrepancy between training and inference phases is known under the name of exposure bias in the machine learning literature [9]. The main issue with TF is that, even if we optimize the average MSE over the h-step horizon in Eq. (2), we are not really doing a multi-step forecasting (we can say that it is a sequence of h single-step forecasting, similarly to what happens in FF-recursive). In other words, the predictor is not specifically trained on a multi-step task.

We thus propose to train the LSTM without TF (LSTM-no-TF) so that the predictor's behavior in training and inference coincides (Fig. 2, right).

3 Forecasting Deterministic Chaos

We initially consider the forecasting of noise-free dynamics derived from some archetypal chaotic systems. The first is the logistic map, a one-dimensional system traditionally used to model the population dynamics:

$$y(t + 1) = r \cdot y(t) \cdot \left(1 - y(t)\right), \tag{4}$$

where the parameter r represents the growth rate at low density. We then consider the Hénon map in its generalized m-dimensional version:

$$y(t + 1) = 1 - a \cdot y(t - m + 2)^2 + b \cdot y(t - m + 1). \tag{5}$$

Two versions of the Hénon map with m equal to 2 and 10 are implemented. As shown in Eqs. (4) and (5), these systems can be easily rewritten as nonlinear autoregressions with m lags.

Figure 3 reports the accuracy of the predictors in terms of R^2-score, with the predictive horizon also rescaled in terms of Lyapunov times, i.e. the inverse of the largest Lyapunov exponent measuring the system's chaoticity (horizontal axis on top). This allows to standardize the performances taking into account the chaoticity of the considered chaotic map.

The LSTM-no-TF predictor provides the best performance (it forecasts with R^2-score higher than 0.9 for 6–7 Lyapunov times), followed by the LSTM-TF and by the FF-recursive. Even if the ranking is the same in all the systems, the distance between the three predictors' performances is case-specific. In particular, the LSTM-TF has almost the same accuracy of LSTM-no-TF for the logistic map, and becomes closer to the FF-recursive predictor for the 10D Hénon. In this system, the LSTM-TF also shows a counterintuitive trend: its accuracy is not monotonically decreasing for increasing values of the lead time as one would expect. This is related to the training

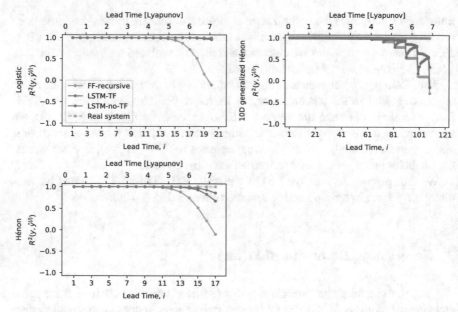

Fig. 3 R^2-score trends across the multi-step horizon for the three considered chaotic systems in a noise-free environment

with TF, which does not allow the predictor to properly propagate the information across time (the issue can be solved by training without TF).

4   Evaluating the Effect of Noise

This chapter extends the analysis to noisy and time-varying environments representing one of the first works, together with [5, 10, 11], that tries to bridge the gap between the ideal deterministic case and the practical applications.

4.1   Observation Noise

The type of noise usually taken into account is that originated by the uncertainty on the observations. Observation noise can be modeled as an additive white disturbance sampled from a gaussian distribution with a null average. Two levels of noise (i.e., two values of the noise standard deviation) are considered: 0.5% and 5% of the noise-free process standard deviation, respectively.

The results of forecasting these noisy time series are reported in Fig. 4 consistently with the deterministic case in Fig. 3. The ranking obtained in both the noise levels is the same as the noise-free one. The most performing predictor, LSTM-no-TF,

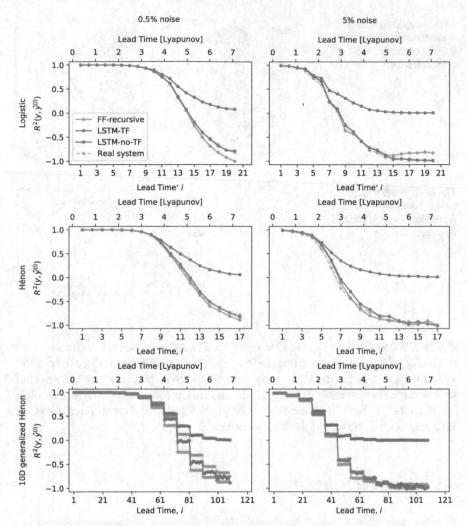

Fig. 4 R^2-score trends across the multi-step horizon for the three chaotic systems. Two different levels of observation noise (0.5 and 5%) are considered

provides a high accuracy (R^2-score >0.9) for about 3 and 1 Lyapunov times with noise levels of 0.5% and 5%, respectively. When the real system is used as predictor, it provides a perfect forecasting in the noise-free case (see Fig. 3), but its accuracy is almost identical to that of the FF-recursive and LSTM-TF predictors in a noisy environment.

Figure 4 also shows that the predictors have two distinct behaviors in the final part of the forecasting horizon. The R^2-scores of FF-recursive and LSTM-TF predictors tend to -1, meaning that the neural networks reproduce fairly well the chaotic attractors' climate. This happens because these two predictors are trained on a single step

Fig. 5 Dataset obtained
from the simulation of the
logistic map affected by
structural noise

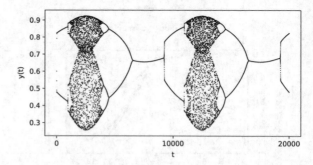

Fig. 6 R^2-score trends
across the multi-step horizon
for the logistic map affected
by structural noise

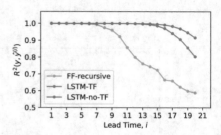

ahead, and thus focus on system identification. The fact that the real systems used as
predictors tend to -1 confirms that the two predictors correctly reproduce the chaotic
systems' regime dynamics. Conversely, the LSTM-no-TF predictor's R^2-scores tend
to 0 because it specifically focuses on the multi-step forecasting. When it reaches
its predictability limit, it becomes a trivial predictor always forecasting the dataset
average value till the end of the h-step horizon.

4.2 Structural Noise

Another type of noise which frequently affects the real-world processes is the so-
called structural noise, i.e., the presence of underlying dynamics that make the considered system non-stationary. We implement the structural noise by periodically
varying the logistic parameter r in Eq. (4). As it happens with the annual periodicity
of the meteorological variables, the variation of r is much slower than that of y. The
resulting slow-fast dynamics showed in Fig. 5 assumes stable, periodic, and chaotic
behaviors, thus representing a challenging testing ground for the neural predictors.

Figure 6 reports the forecasting accuracy obtained with the three predictors. The
wide gap between the two LSTM nets and the FF-recursive predictor indicates that
a recurrent structure is more appropriate than a static one for the considered non-
stationary task. Once again, a training without TF further increases the predictive
accuracy of the LSTM predictor.

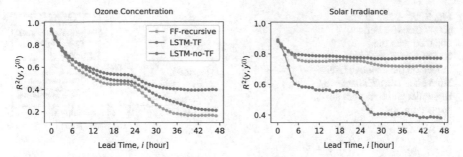

Fig. 7 R^2-score trends across the 48-h horizon for the ozone concentration (left) and solar irradiance (right) datasets

5 Real-World Applications

In the end, we evaluated the performances of the three neural predictors on two real-world time series, which are affected by both observation and structural noise (i.e., the uncertainty on the measurement and the presence of yearly and daily periodic dynamics).

5.1 Ozone Concentration

The first practical application considers a dataset of ground-level ozone concentration recorded in Chiavenna (Italy). The time series covers the period from 2008 to 2017 and is sampled at an hourly time step. The physical and chemical processes which generates the tropospheric ozone are strongly nonlinear and its dynamics is chaotic, as demonstrated by the positive value we estimated for the largest Lyapunov exponent (0.057).

The results obtained on a 48-h predictive horizon are reported in Fig. 7 (left). The predictive accuracy of the three predictors is almost identical in the first 6 h. After that, a relevant gap emerges confirming the rank of the previous numerical experiments: LSTM-no-TF ensures the best performance, followed by the LSTM-TF and the FF-recursive predictors.

5.2 Solar Irradiance

The second real-world case study is an hourly time series of solar irradiance, recorded at Como (Italy) from 2014 to 2019. Its largest Lyapunov exponent is equal to 0.062, indicating that also its dynamic is chaotic.

Fig. 8 R^2-score trends across the 48-h horizon obtained with the LSTM-no-TF predictor trained on the Como dataset. Source domain). Casatenovo, Bigarello, and Bema are the target domains

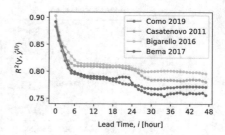

The R^2-scores of the predictors reported in Fig. 7 (right) show that the FF-recursive predictor provides almost the same accuracy of the LSTM-no-TF, especially in the first 24 h of the horizon. Unlike in the other cases, the LSTM-TF predictor performance is so low that it is practically useless after few hours ahead (for comparison, the clear-sky model has R^2-score equal to 0.58).

We also use the solar irradiance dataset to test the generalization capability of the neural predictors in terms of domain adaptation. The LSTM-no-TF predictor trained on the data recorded at the Como station, is used, without retraining, to forecast the solar irradiance in three other sites with quite heterogeneous geographical conditions and in different years: Casatenovo (2011), Bigarello (2016), and Bema (2017). Figure 8 reports the results of the domain-adaptation analysis. The R^2-scores obtained are almost identical in the four domains (note that the vertical axis has been zoomed to spot the small differences between the four trends). This is an interesting result since it tells us that we can identify a neural predictor with a (hopefully large) dataset recorded at a certain location, and then use it in another location where an appropriate dataset for training is not available.

6 Conclusion

We tackled the problem of chaotic dynamics forecasting by means of artificial neural networks, implementing both feed-forward and recurrent architectures. A wide range of numerical experiments have been performed. First, we forecasted the output of some archetypal chaotic processes in a deterministic environment. The analysis is then extended to noisy dynamics, taking into account both observation (white additive disturbance) and structural (non-stationarity) noise. Finally, we considered two real-world time series of ozone concentration and solar irradiance, which have chaotic behaviors.

Whatever the system or the type of noise, our results showed that LSTM-no-TF is the most performing multi-step predictor. The distance in the performances obtained with LSTM-no-TF and the other predictors (FF-recursive and LSTM-TF) is essentially task dependent.

The better predictive power of the LSTM-no-TF predictor is due to the fact that it is specifically trained for a multi-step forecasting task. Conversely, the FF-recursive and LSTM-TF predictors are optimized solving a system identification task. They are somehow more generic models, suitable for the mid-short-term forecasting and also able to replicate the long-term climate of the chaotic attractor (they can be used, for instance, to perform statistical analyses and for synthetic time series generation).

Besides the accuracy of the forecasting, the domain-adaptation analysis attested the high generalization power of the neural predictors across a relatively heterogeneous domain. The techniques presented in this chapter are specifically developed for multi-step forecasting. Therefore, they could be particularly interesting in the context of receding-horizon control schemes such as model predictive control.

Acknowledgements The author would like to thank Prof. Giorgio Guariso and Prof. Fabio Dercole that supervised this research and offered deep insights into this study.

References

1. M. Sangiorgio, S. Barindelli, R. Biondi, E. Solazzo, E. Realini, G. Venuti, G. Guariso, Improved extreme rainfall events forecasting using neural networks and water vapor measures, in *6th International conference on Time Series and Forecasting*, pp. 820–826 (2019)
2. M. Sangiorgio, S. Barindelli, V. Guglieri, R. Biondi, E. Solazzo, E. Realini, G. Venuti, G. Guariso, A comparative study on machine learning techniques for intense convective rainfall events forecasting, in *Theory and Applications of Time Series Analysis* (Cham, 2020. Springer), pp. 305–317
3. F. Dercole, M. Sangiorgio, Y. Schmirander, An empirical assessment of the universality of ANNs to predict oscillatory time series. IFAC-PapersOnLine **53**(2), 1255–1260 (2020)
4. M. Sangiorgio, Deep learning in multi-step forecasting of chaotic dynamics. Ph.D. thesis, Department of Electronics, Information and Bioengineering, Politecnico di Milano (2021)
5. D. Patel, D. Canaday, M. Girvan, A. Pomerance, E. Ott, Using machine learning to predict statistical properties of non-stationary dynamical processes: system climate, regime transitions, and the effect of stochasticity. Chaos **31**(3), 033149 (2021)
6. J. Pathak, Z. Lu, B.R. Hunt, M. Girvan, E. Ott, Using machine learning to replicate chaotic attractors and calculate lyapunov exponents from data. Chaos **27**(12), 121102 (2017)
7. G. Guariso, G. Nunnari, M. Sangiorgio, Multi-step solar irradiance forecasting and domain adaptation of deep neural networks. Energies **13**(15), 3987 (2020)
8. M. Sangiorgio, F. Dercole, Robustness of LSTM neural networks for multi-step forecasting of chaotic time series. Chaos Solitons Fractals **139**, 110045 (2020)
9. T. He, J. Zhang, Z. Zhou, J. Glass, Quantifying exposure bias for neural language generation (2019). *arXiv preprint* arXiv:1905.10617
10. J. Brajard, A. Carrassi, M. Bocquet, L. Bertino, Combining data assimilation and machine learning to emulate a dynamical model from sparse and noisy observations: a case study with the Lorenz 96 model. J. Comput. Sci. **44**, 101171 (2020)
11. P. Chen, R. Liu, K. Aihara, L. Chen, Autoreservoir computing for multistep ahead prediction based on the spatiotemporal information transformation. Nat. Commun. **11**(1), 1–15 (2020)

Optimal Management and Control of Smart Thermal-Energy Grids

Stefano Spinelli⊙

Abstract This work deals with the development of novel algorithms and methodologies for the optimal management and control of thermal and electrical energy units operating in a networked configuration. The aim of the work is to foster the creation of a smart thermal-energy grid (smart-TEG), by providing supporting tools for the modeling of subsystems and their optimal control and coordination. A hierarchical scheme is proposed to optimally address the management and control issues of the smart-TEG. Different methods are adopted to deal with the features of the specific generation units involved, e.g., multi-rate MPC approaches, or linear parameter-varying strategies for managing subsystem nonlinearity. An advanced scheme based on ensemble model is also conceived for a network of homogeneous units operating in parallel. Moreover, a distributed optimization algorithm for the high-level unit commitment problem is proposed to provide a robust, flexible and scalable scheme.

Keywords Hierarchical optimization and control · Model predictive control · Energy systems

1 Introduction

The ecological and energetic transition is driving a deep transformation of the energy and utility industry, promoting a shift from centralized to distributed generation systems. This transformation is supported by advanced digital technologies and requires the introduction of new paradigms for energy management and control, including a radical change in the user role, advancing the definition of the control problems from the device level to a system level, and by the integration of different energy vectors.

A paradigm shift in the consumers' function is promoted by the decentralization of the energy production and the integration of small scale generation, giving them

S. Spinelli (✉)
Dipartimento di Elettronica, Informazione e Bioingegneria, Politecnico di Milano, Via Ponzio 34/5, 20133 Milano, Italy

STIIMA, Consiglio Nazionale delle Ricerche, Via A. Corti 12, 20133 Milano, Italy
e-mail: stefano.spinelli@polimi.it; stefano.spinelli@stiima.cnr.it

© The Author(s) 2022
L. Piroddi (ed.), *Special Topics in Information Technology*,
PoliMI SpringerBriefs, https://doi.org/10.1007/978-3-030-85918-3_2

an active role by becoming local producers, as well as by modifying proactively their consumption patterns. In this new scenario, the consumer is asked to abandon the perspective of "Energy as a Commodity", where energy is considered as always available and cost-effective on demand, toward the paradigm of "Energy-as-a-Service". This new paradigm requires to move the focus of management, optimization, and control problems from the device-level to the system level, e.g., as in smart grids, promoting a scenario in which the production of various forms of utilities is more and more integrated. A step-ahead in system flexibility is built indeed on a cross-sectorial integration of different energy vectors and on the development of tools and technologies that enable efficient utilization of multi-dimensional energy systems. For this reason, significant research efforts must be devoted to alternative energy carriers, e.g., compressed air, heating and cooling networks, and to their integration. The concept of *energy hub* fosters efficiency improvement for multiple energy carrier systems through analysis tools, dynamic modeling, and the control of interacting subsystems [1]. As reported in the review [2], the optimal management and control of energy hubs is one of the most important open issues.

This requires to extend the idea of smart grids to *Smart Thermal Energy Grid* (Smart-TEG): the generation nodes are composed of multiple integrated units that require coordination and control to fulfill the demand of the various consumers, which in turn may vary (in part) their demand based on optimization criteria.

2 The Smart-TEG Problem

In this work we focus on smart-TEG in industrial scenarios. A typical case is represented by industrial parks: their energy-centers generate electrical power and thermal energy that are distributed across the site to companies. Some of these, at the same time, are provided with on-site generation units (GUs) to guarantee a degree of autonomy. In this scenario, local generation units and the industrial-park energy facilities can be combined together to improve the overall efficiency of the network of subsystems, managing and controlling these units in a wider context, i.e., as nodes of a more extended grid.

The thermal and electricity demand defined by the production scheduling must be fulfilled at any time, however the mix between on-site generated energy and net exchange with the main grid must be optimized, as well as the commitment of the GUs and their operating point. The control of the integrated generation units become a necessary element to operate them safely and efficiently.

While the eco-design, retrofitting, and revamping of energy system produce the greatest improvement on the efficiency, an enhanced management and control of existing GUs can provide beneficial results with a lower capital investment. By exploiting the integration of the multi-generation systems on the plant and by extending the grid outside company boundaries the overall efficiency can be improved greatly. Therefore, the focus has to be moved on the synergies among various forms of energy. The presence of various energy carriers, the complexity of interconnections at different levels among the subsystems require the development of intelligent technologies for its management and control, e.g., by inheriting and transposing solutions from the smart grid context.

Most of the ideas and approaches discussed in this work can be extended to a broader class of problems of industrial and scientific interest e.g., multi-line (parallel) plants with raw material or energy/power constraints; water/steam supplier (consumer) networks.

Despite the relevance of these problems, many issues are still present regarding coordination and control of these systems. Classical centralized solutions are limited by computational issues, lack of scalability and privacy concerns. The industrial need for an extension of dynamic optimal management strategies, from subsystem to plant-wide level, is well known: new strategies and methods must be put in place to approach plant-wide optimal management.

3 Optimization-Based Hierarchical Control Solutions

To address the optimal control and management of a smart-TEG, a hierarchical approach is here adopted. The overall problem is decomposed by considering the temporal separation that exists among the system dynamics, and the different horizon of the problems at various levels. Several sub-problems can be identified to achieve the overall target: (i) the unit commitment (UC) and scheduling of the generation units; (ii) the optimization of the operating point for each unit; (iii) the dynamic coordination of interacting units; (iv) the regulation and control of the single unit. The interaction among the different layers is taken into account to provide an integrated solution.

Model-based optimization and control is assumed as a common framework for the development of the solution strategies at the different layers: model predictive control (MPC) and, in general, receding horizon formulations are used.

The architecture for the smart-TEG management considers a multi-layer approach. In the top layer, the UC problem and the economic dispatch optimization is dynamically solved, considering networked systems sharing a set of common resources. The optimization at this level can benefit from decomposition and parallelization strategies, based on distributed and decentralized approaches, see Sect. 3.3.

At the lower layers, hierarchical and distributed control levels based on MPC are defined: advanced control solutions for single generation units are developed to enable the integration with the upper layers, and a control scheme for the coordination of homogeneous ensemble of subsystems is also proposed. The latter exploits subsystem configuration to reach a top-down integration, from scheduling to dynamical control, which is scalable and flexible.

At any level, the proper modeling of the systems becomes a fundamental and critical task: an extensive discussion is carried out in [3] [Part II], whereas a brief digression on hybrid models for UC optimization is given in the following section.

3.1 Unit Commitment and Control of Generation Units

The coordination of the generation units in a node of the smart-TEG, i.e., inside a company, is itself a local problem of UC and control for the set of interacting subsystems. A two-level hierarchical scheme is here proposed. The hierarchical control structure, see Fig. 1, includes a high layer that aims to optimize the performance of the integrated plant on a day-ahead basis with a longer sampling time (typically 15 min.) and a low-level regulator to track the set-points. The model used at the high level is sketched in Fig. 2.

For a proper management of the UC it is necessary to extend the mathematical description of the system to include not just the production modes, but also start-up and shut-down dynamics, as well as any other particular operating mode. This high-level model is devised as a discrete hybrid automaton [4], combining discrete and continuous states and inputs, that permits to merge the finite state machine describing the operating modes of the unit and their transitions, with the continuous dynamics of the system inside each mode. Here the transition is defined by a combination of continuous state dynamics, describing the mode dwell times, and exogenous Boolean commands.

The high-level optimizer determines the future modes of operation of the systems, as well as the optimal production profile for the whole future optimization. The receding horizon optimization is formulated as a mixed-integer linear program (MILP), due to the presence of both continuous and discrete decision variables, a linear objective function defining the operating cost, and linear inequalities enforcing demand satisfaction.

The lower level is a dynamic MPC that operates on each individual system to guarantee that the process constraints are fulfilled, with a faster sampling time. For the dynamic control of the sub-system, specific controllers can be designed for the different operating modes: a linear MPC is implemented for normal production modes (see [5]; or [6] where multi-rate approach is adopted), while to address the nonlinearity of the system in the start-up phase, a linear parameter-varying MPC approach [5] is proposed for the optimization of a nonlinear system subjected to hard constraints. This method, exploiting the linearization of the system along the predicted trajec-

Fig. 1 Hierarchical scheme for unit commitment and control

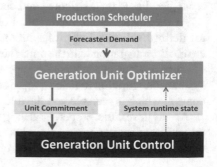

Fig. 2 Operating modes of
interacting GUs

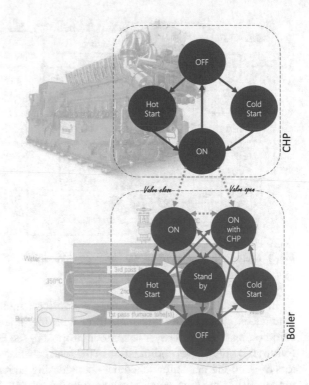

tory, is able to address nonlinear system with the advantage of the computational
time with respect to NMPC strategy, as an approximation of the SQP optimization
stopped at the first iteration. The scheme has been validated on an industrial use-case.
An extensive presentation is reported in [5].

3.2 An Ensemble-Model Approach for Homogeneous Units

Energy plants are typically characterized by a number of similar subsystems operating in parallel: this ensures reliability, continuity of service in case of unexpected
breakdowns or planned maintenance, and modularity of operation. In particular,
when the demand fluctuation range is wide, a unique GU is forced to operate far
from its most efficient working conditions: so, instead of operating at inefficient low
regimes at low demand, a set of cooperative units can be opportunely committed to
run in the most efficient configuration.

These units are characterized to be *homogeneous*, i.e., similar but not identical, with the same types of inputs and outputs but slightly different parameters and
dynamics: e.g., GUs of different producers or product generations.

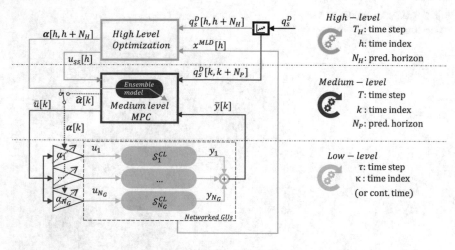

Fig. 3 Multi-layer solution for parallel homogeneous units

In this context, the aim is to design a solution which is both robust and flexible, guaranteeing modularity and scalability with the objective of jointly sustaining a common load. A multi-layer scheme is proposed, see Fig. 3, and it is composed as follows: (i) a high-level optimization of the load partition—defined by the sharing factors α—and the generator schedule—considering activation dynamics by a hybrid model; (ii) at medium level, a robust tube-based MPC that tracks a time-varying demand using a centralized, but aggregate, model—the *ensemble* model—whose order does not scale with the number of subsystems; (iii) at low-level, decentralized controllers to stabilize the generators. In Fig. 3, the decentralized lowest-level controllers are not shown and the closed-loop nonlinear systems are indicated by $\mathcal{S}_i^{CL}$. The multi-rate nature of the scheme is highlighted in the figure by the presence of different time indices (h, k, κ) respectively for the high/medium/low layers, discretized with time step $(T_H > T > \tau)$. Without delving too much into technicalities, that can be found in [7, 8], three main elements characterize this control solution: (i) the *sharing factors* to partition load and inputs to the subsystems; (ii) the *ensemble model* to manage the overall system in a scalable way; (iii) the *robust medium-level MPC* to manage intrinsic modeling mismatch.

The sharing factors. To partitioning the demand to the N_G subsystems that constitute the plant, the sharing factors α_i are introduced. These are characterized by the following properties: $0 \leq \alpha_i \leq 1$ and $\sum_{i=1}^{N_G} \alpha_i = 1$. Thus, given the overall input of the medium level, $\bar{u}$, the local input to the generic subsystem i can be defined by $u_i = \alpha_i \bar{u}$. The sharing factors are computed in receding horizon by the top-layer optimizer, based on the forecast of the demand.

The ensemble model. The sharing factors are essential in the derivation of the global model of the ensemble. For assumption, each controlled sub-system can be described by a state-space linear model. To compose the ensemble model, it is required to define an artificial *reference* model, which has the same state vector

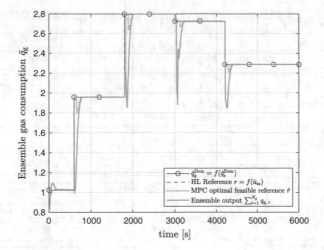

Fig. 4 Ensemble total output and demand

for each subsystem. While homogeneity implies by definition coherent inputs and outputs, the state vectors might differ: thus, reference-model state might be defined by a proper transformation. By design reference-model the state and output matrices are the same for each sub-system. Instead, the input matrix is derived by enforcing gain consistency conditions [9], which guarantee the same static gain of the original system. It has to be remarked that during transient the gain consistency does not hold and, due to model mismatch, a disturbance term must be introduced. Based on these reference models, the ensemble model can be defined considering the parallel configuration and the sharing factor definition.

The robust medium-level MPC. Due to the presence of this disturbance term, a *robust* medium-level MPC must be designed to dynamically controls the overall input to follow the time-varying demand. A tube-based MPC approach [10] is used: based on the ensemble model, it can be dynamically modified by the high-level optimization following sharing factors trajectories [7]. By enforcing both local and global constraints, we can guarantee that also each sub-system respects local bounds. In specific conditions, e.g., with sharp demand variations, abrupt changes of ensemble configuration may lead to medium-level MPC infeasibility: a nonlinear optimization addresses MPC infeasibility considering the sharing factors as temporary decision variables and driving the system to the new required configuration via a feasible path.

The results for a set of 5 water-tubes boilers operating in parallel are here reported: the Figs. 4 and 5 show how the medium level MPC can track the gas demand both globally and locally, while in Figs. 6 and 7 the optimal unit commitment, i.e. operating mode evolution, and the best sharing factors are shown. In Fig. 7 the transitional configuration computed by the medium-level nonlinear optimization are also reported, compared with high-level optimization.

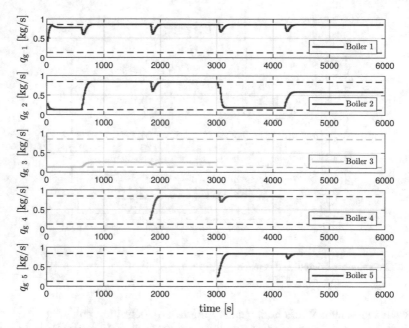

Fig. 5 Local output and demand

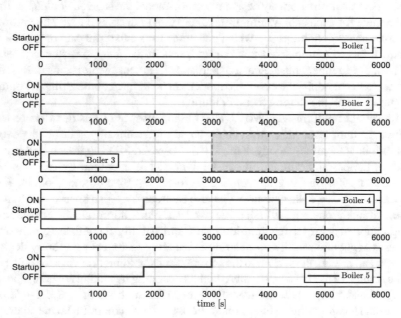

Fig. 6 Operating modes

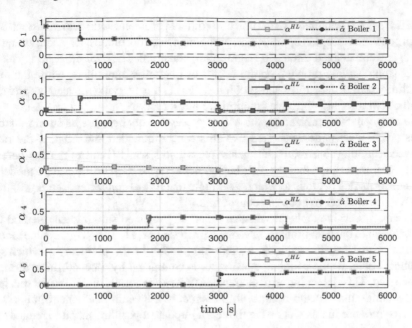

Fig. 7 Sharing factors

Note that this control scheme enables the dynamical modification of the ensemble configuration and plug and play operations.

3.3 A Distributed Unit Commitment Optimization

The UC optimization aims to compute the best operating strategy for coordinating the subsystems. We explicitly consider the smart-TEG as composed of different companies owning and controlling a subset of subsystems, that must cooperate to guarantee the fulfillment of a global demand, while restricting the number and type of information communicated across the network. A centralized optimization is most of all limited by flexibility and robustness to network variations, and privacy concerns. The centralized formulation, in fact, implies that all the companies participating in the network provide to the central computation unit their cost function and their constraint sets. This correspond to actually disclose their preferences and limitations, as well as the models of the internal GU and the company KPIs. To overcome these issues, a distributed multi-agent optimization is here proposed. A peculiarity of smart-TEG is that the network of energy generation agent is characterized by different hierarchical levels, see Fig. 8, based on the ability to reach (almost) all the consumers or just a local cluster of them, and defined by the piping network and steam flow direction. Thus, the units are classified as either Central GUs or Local GUs.

A MILP can be formulated for the unit commitment problem, as discussed in Sect. 3.1. Such constraint-coupled optimization problems can be decomposed by addressing its dual problem, defined through a Lagrangian relaxation. The dual program is indeed a decision coupling problem on the vector of Lagrange multipliers, while the dual function exhibits decoupled sub-problems, see [11], preserving the agents' privacy: local constraints and objective function are not shared.

As a non-convex optimization, the MILP presents a duality gap, as strong duality does not hold. It is demonstrated that this can be bounded and small if the ratio between coupling constraints and agents is less than 1, see [12] and references therein. Therefore, we assume that the distribution network for each utility can be modeled as an informatics fieldbus to which the single agents are connected, see Fig. 8, thus limiting the number of coupling constraints.

Decentralized hierarchical schemes based on projected sub-gradient method for the solution to the dual problem have been proposed in [11, 12]. To guarantee recovering a primal feasible solution, a modified version of the problem, in which the resource vector of the coupling constraints is contracted by a penalization term, is considered. This tightening term guarantees finite-time feasibility. However this feasible solution may be sub-optimal with respect to the centralized one. Similarly to [12], we propose in this work a the tightening update algorithm aimed to reduce this sub-optimality.

A fully distributed approach can improve the privacy preservation by limiting the information exchanged among the network [13]. To achieve the full distribution, a

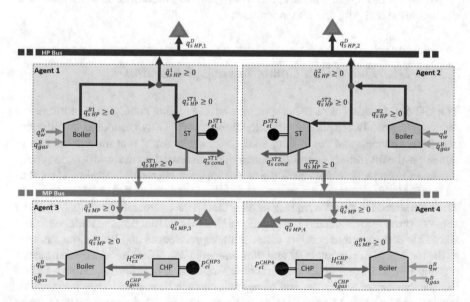

Fig. 8 Central and Local GUs with bus assumption. On top (with red background) Central GUs, at the bottom Local GUs (orange). Different subsystems are present in each node (Boilers, Steam Turbines (ST), CHP). Consumers are depicted as triangles

local copy of the vectors of Lagrange multipliers and of the penalization vector is introduced: only these local estimations are shared across the network by a peer-to-peer communication. The algorithm in [13] guarantees feasibility, leaving an open issue regarding solution sub-optimality. In this work, supported by this finite-time feasibility guarantee, we propose a distributed algorithm with performance focus, that can reach an optimal solution close to centralized method.

To permit an improvement of the global cost, we are forced to relax slightly the privacy concerns, playing with the intrinsic trade-off that exists between information sharing and optimality. The agents are obliged to disclose their local contribution to the coupling constraint and the local cost to their neighbors. Thanks to distributed average consensus routines, that permit the estimation of the overall feasibility and cost improvement, each agent achieves a "common knowledge" from which it is impossible to recognize and determine the specific contribution of any other agent, thus preserving privacy. Rather, this is essential to empower each agent to evaluate autonomously the feasibility and optimality of the current iteration, and take the opportune actions.

Both the proposed decentralized and distributed algorithm show an improvement in the global cost of the smart-TEG with respect to state of the art approaches. A possible drawback of the proposed distributed algorithm is related to a highly intense—but local—data exchange requirement to achieve an enhancement in optimality performance.

Detailed discussion about the algorithms and results can be found in [3].

4 Conclusions

This chapter aimed to give an overview of the hierarchical architecture for the management and control of smart-TEG. This research work has addressed the solutions to enable efficient utilization of multi-dimensional energy systems, developing control algorithms, tools, and technologies at different levels. The multi-layer approach presented in Sect. 3.1 is shown to be effective in optimizing the scheduling and the unit commitment of a set of interacting generation units, by exploiting the hybrid model for representing all the operating modes, but also for the dynamic regulation of each subsystem. Moreover, when homogeneous units are present, as discussed in Sect. 3.2, the ensemble-model approach can guarantee a scalable solution, with plug&play features. In addition, decentralized and distributed algorithms for the high-level unit commitment problem with performance focus have been introduced.

The current hierarchical approach is designed to consider the low-level controllers as decentralized: a possible future extension of the scheme can benefit by replacing the lower layer with a distributed cooperative MPC approach, in which the different subsystems can exchange with the others their predicted solutions. Future work will consider also the improvement of the multi-layer scheme for homogeneous units by comparing the overall performance with the implementation of an additional low-level shrinking MPC control to further address the local model mismatch. We also envision to extend the high level optimization including also the ensemble dynamics. Currently the ensemble algorithm is designed for linear models at the lower levels, an extension to nonlinear system is foreseen.

References

1. M. Geidl, G. Koeppel, P. Favre-Perrod, B. Klockl, G. Andersson, K. Frohlich, Energy hubs for the future. IEEE Power Energy Mag. **5**(1), 24–30 (2006)
2. M. Mohammadi, Y. Noorollahi, B. Mohammadi-ivatloo, M. Hosseinzadeh, H. Yousefi, S.T. Khorasani, Optimal management of energy hubs and smart energy hubs—a review. Renew. Sustain. Energy Rev. **89**, 33–50 (2018)
3. S. Spinelli, Optimization and control of smart thermal-energy grids. Doctoral Thesis, Politecnico di Milano, pp. 1–320 (2021)
4. H. Lin, P. Antsaklis, N. Publishers, Hybrid Dynamical Systems: An Introduction to Control and Verification. Found. Trends Sys Control. **1**(1), 1–172 (2014). https://doi.org/10.1561/2600000001
5. S. Spinelli, M. Farina, A. Ballarino, An optimal hierarchical control scheme for smart generation units: an application to combined steam and electricity generation. J. Process Control **94**, 58–74 (2020)
6. X. Zhang, M. Farina, S. Spinelli, R. Scattolini, Multi-rate model predictive control algorithm for systems with fast-slow dynamics. IET Control Theor. Appl. **12**(18), 2468–2477 (2018)
7. S. Spinelli, M. Farina, A. Ballarino, A hierarchical architecture for optimal unit commitment and control of an ensemble of steam generators. IEEE Trans. Control Syst. Technol. 1–14 (2021). https://doi.org/10.1109/TCST.2021.3094886
8. S. Spinelli, E. Longoni, M. Farina, F. Petzke, S. Streif, A. Ballarino, "A hierarchical architecture for the coordination of an ensemble of steam generators." IFAC-PapersOnLine **53**(2), 557–562 (2020)
9. F. Petzke, M. Farina, S. Streif, A multirate hierarchical MPC scheme for ensemble systems, in *Proceedings of the Conference on Decision and Control* (2018), pp. 5874–5879
10. G. Betti, M. Farina, R. Scattolini, A robust MPC algorithm for offset-free tracking of constant reference signals. IEEE Trans. Autom. Control **58**(9), 2394–2400 (2013)
11. R. Vujanic, P.M. Esfahani, P.J. Goulart, S. Mariéthoz, M. Morari, A decomposition method for large scale MILPs, with performance guarantees and a power system application. Automatica **67**, 144–156 (2016)
12. A. Falsone, K. Margellos, M. Prandini, "A decentralized approach to multi-agent MILPs: finite-time feasibility and performance guarantees." Automatica **103**, 141–150 (2019)
13. A. Falsone, K. Margellos, M. Prandini, A distributed iterative algorithm for multi-agent MILPs: finite-time feasibility and performance characterization. IEEE Control Syst. Lett. **2**(4), 563–568 (2018)

Electronics

Application Specific Integrated Circuits for High Resolution X and Gamma Ray Semiconductor Detectors

Filippo Mele

Abstract The increasing demand for performance improvements in radiation detectors, driven by cutting-edge research in nuclear physics, astrophysics and medical imaging, is causing not only a proliferation in the variety of the radiation sensors, but also a growing necessity of tailored solutions for the front-end readout electronics. Within this work, novel solutions for application specific integrated circuits (ASICs) adopted in high-resolution X and γ ray spectroscopy applications are studied. In the first part of this work, an ultra-low noise charge sensitive amplifier (CSA) is presented, with specific focus on sub-microsecond filtering, addressing the growing interest in high-luminosity experiments. The CSA demonstrated excellent results with Silicon Drift Detectors (SDDs), and with room temperature Cadmium-Telluride (CdTe) detectors, recording a state-of-the-art noise performance. The integration of the CSA within two full-custom radiation detection instruments realized for the ELETTRA (Trieste, Italy) and SESAME (Allan, Jordan) synchrotrons is also presented. In the second part of this work, an ASIC constellation designed for X-Gamma imaging spectrometer (XGIS) onboard of the THESEUS space mission is described. The presented readout ASIC has a highly customized distributed architecture, and integrates a complete on-chip signal filtering, acquisition and digitization with an ultra-low power consumption.

3.1 Introduction

Semiconductor radiation detectors (SRDs) are indispensable elements in a large variety of scientific, industrial and medical instruments. Since 1980 they have experienced a rapid development not only as part of large-scale high-energy physics experiments and in space missions aimed at universe observation for astrophysics studies, but also in multidisciplinary facilities, such as synchrotron light sources, pushing the frontier of the research in a multitude of areas, especially in material and

F. Mele (✉)
Dipartimento di Elettronica, Informazione e Bioingegneria, Politecnico di Milano,
Via Ponzio 34/5, 20133 Milano, Italy
e-mail: filippo.mele@polimi.it

© The Author(s) 2022
L. Piroddi (ed.), *Special Topics in Information Technology*,
PoliMI SpringerBriefs, https://doi.org/10.1007/978-3-030-85918-3_3

life sciences. Despite the diversification of the design focus required in each application, one can always identify three main functional blocks in a radiation detection system; the *sensor*, or the detector, is the first element, where the radiation physically undergoes an interaction that will produce an electric charge Q proportional to the photon deposited energy. This charge is collected by the *front-end electronics* (FEE) or readout electronics, and converted into a voltage or a current signal, which constitutes the second functional block in the radiation detection chain; since its invention in 1955 [1], the charge sensitive amplifier (CSA) is the most used solution for the realization of the first amplification stage of charge signals generated by the sensor, which basically consists of a low-noise amplifier operated with a capacitive negative feedback. Finally the *back-end electronics* (BEE) is responsible for the additional signal processing, digitization and data transmission of the desired information (energy, time-stamping, position of interaction, etc.), which completes the application. By all means, the border which separates the FEE from the BEE can be subject to different interpretations, according to the specific design cases. Within this text we will refer to the BEE as the electronics that can be considered to be more robust to noise issues—thanks to the beneficial processing of the FEE—and typically more relaxed in terms of area occupation, signal routing and power consumption, so that the BEE will be more likely to be placed further from the detection plane.

When in 1984 Emilio Gatti and Pavel Rehak proposed a novel ingenious charge transport mechanism for silicon detectors, which took the name of semiconductor drift chambers or silicon drift detectors (SDD) [2], a revolution in the performance of semiconductor detectors started; these detectors allowed to obtain the same spatial resolution of microstrip detectors using a much reduced number of preamplifiers, showed an architecture which could make the detector's capacitance independent of the detector's active area, with an unprecedented improvement in the noise performances, and was based on the planar technology, which, at the time, was already well consolidated. Due to the delicate inter-operation between the sensor and the FEE, this innovation leap pushed the development of tailored readout electronics and, with the advent of integrated circuits (IC), a new world of possibilities in terms of customization and integration for both the FEE and BEE, whose extreme performances in nuclear microelectronics are still topic of research and investigation.

Within this work the latest results of the state of the art ICs for low capacitance (<0.2 pF) X and γ ray detectors are presented. In the first part of this chapter, an ultra-low noise CSA for synchrotron radiation facilities will be shown: the paramount requirement of a high-resolution performance, and the relaxed constraints on the power budget and area occupation, represent a golden opportunity to realize a preamplifier with state-of-the-art characteristics in terms of noise and speed. On the other hand, in the second part, the design of a complete readout chain of an X/γ imaging spectrometer for astrophysics radiation instrumentation will be described, which is an example of a highly customized application specific integrated circuit (ASIC) with a complex multi-channel distributed architecture.

3.2 SIRIO-6: A New Generation of CSAs for High-Rate, High-Resolution X-/γ-ray Spectroscopy

The quest for a minimum noise solution for the amplification of charge signals produced by SRDs mostly relies on the correct choice of a front-end device, capable of assuring a high gain and a low noise.[1] For CSAs realized in CMOS technology, proposed for the first time in 1997 [3], the optimal noise performance can be reached at the so-called capacitive matching condition on the total gate capacitance, which has a different expression according to the dominant noise component [4], that, in turn, is subject to the desired (or available) processing time in the acquisition chain. Among the main challenges in the realization of CSAs, the balance between the noise performance, and the processing speed is surely a topic of recent interest; indeed, new generation detection systems are often required to withstand high input photon rates (up to several million-counts per second), keeping an adequate spectroscopic resolution. In the framework of the realization of two new custom radiation detection systems for the Elettra synchrotron in Trieste (Italy) and the SESAME synchrotron in Allan (Jordan) a new family of fast, pulsed-reset preamplifiers, named SIRIO-6, has been developed. The main aim of this project is to move the design-focus toward sub-microsecond processing times, allowing such research-grade systems to move toward the million-counts per second range (Mcps). In this section the SIRIO-6 experimental characterization will be presented showrly presented, together with the integration on the complete detection systems at the Elettra and SESAME synchrotrons.

3.2.1 SIRIO for High Resolution Silicon Drift Detectors

Being optimized for low capacitance SRDs, SIRIO-6 finds its ideal application in combination with SDDs, that, due to their excellent noise performance at non-cryogenic temperatures, are among the most widely used devices in energy dispersive spectroscopy (EDS). With respect to previous designs, SIRIO-6 has been concieved with the specific goal of optimizing the noise performance at fast processing times—generally identified in terms of the peaking-time (τ_{peak}) of the following filtering stages—and to have a rapid transient response and low reset dead-time [5]. In Fig. 3.1 a comparison between the spectroscopic resolution on the ^{55}Fe 5.9 keV line using the new SIRIO CSA with respect to a previous generation is shown at different peaking times, when connected to a $10\,mm^2$ SDD with similar leakage performance. The measurement, reported at a moderate cooling of $-35\,°C$ shows a drastic improvement in the noise performance, especially at the lowest peaking times, with an excellent full width half-maximum (FWHM) of 138.4 eV on the Mn-

---

[1] The noise figure (also known as the excess noise ratio) is historically the most common quality factor for low-noise amplifier designs. Nevertheless, this parameter, whose definition is based on the assumption of a conventional fixed noise at the input source, is not suitable for the description of CSA, which are rather characterized by their equivalent noise charge (ENC).

Fig. 3.1 In the upper part, the 5.9 keV Mn-K$_\alpha$ FWHM measured with the new SIRIO-6 (in red) with respect to the best performance of the previous SIRIO generation (in blue), using a silicon drift detector at moderate cooling ($-35\,°C$). The bottom lines represent the estimated electronic noise as obtained by subtracting the Fano noise contribution

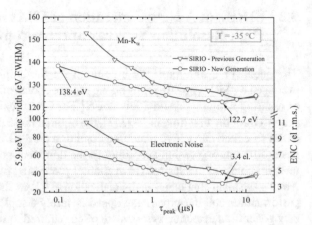

K$_\alpha$ line at the peaking time of 0.1 μs, corresponding to an ENC of 8.1 el. r.m.s., which is among the best found in literature for such a fast shaping at the time of writing [6]. The optimum resolution is achieved at 5.6 μs, where a FWHM of 122.7 eV at 5.9 keV is recorded, corresponding to an electronic noise of 3.4 el. r.m.s. (29.9 eV FWHM).

3.2.2   SIRIO for CdTe Detectors at Deep Sub-microsecond Signal Processing Times

Due to their low detection efficiency, silicon detectors are not suited for high-energy photons (>20 keV). With this respect, thanks to their excellent absorption efficiency and their wide bandgap voltage which allows room temperature operation, CdTe and CdZnTe detectors are attracting a growing interest in the scientific and industrial community. The SIRIO-6 CSA specifications well match the needs of these kind of compound semiconductor detectors, especially at high fluxes of radiations, which are required to minimize the acquisition time in high-throughput X-/γ-ray scanners used for medical diagnostic, quality control and homeland security. Using SIRIO-6 and a custom array of CdTe pixels, we were able to achieve at $\tau_{peak} = 50$ ns an electronic noise of around 37.6 el. r.m.s., with negligible contribution from the leakage current shot noise. The optimum performance is obtained at 1 μs where an electronic noise of approximately 19.7 el. r.m.s. allows the acquisition of an unprecedented high-resolution spectrum for CdTe detectors, shown in Fig. 3.2: on the 59.5 keV line the resolution of ∼0.78% recorded using the new SIRIO-6 preamplifier improves by a factor of two the state-of-the-art performance previously reported in literature under similar experimental conditions [7]. The usage of the SIRIO-6 CSA with the described detector array demonstrated the feasibility of high-resolution spectroscopy with CdTe detectors at room temperature and at deep sub-microsecond processing

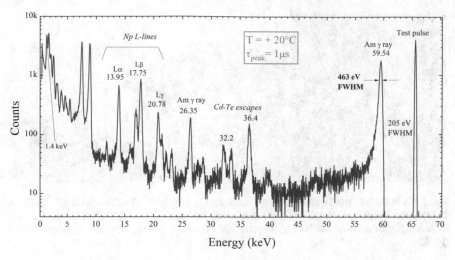

Fig. 3.2 The ^{241}Am spectrum acquired using SIRIO-6 coupled with the CdTe detector for the optimum peaking time (1 μs) at room temperature. The 463 eV FWHM on the 59.5 keV line improves significantly the state-of-the-art resolution for CdTe detectors

times, potentially opening the path for new applications of CdTe detectors in those systems requiring a high detection efficiency and high count-rate capability in the soft-γ energy range, without sacrificing the spectroscopic resolution.

3.2.3 Detection Systems for the Elettra and SESAME Synchrotrons

As part of the INFN-ReDSoX project, the SIRIO CSA is currently being integrated into two newly designed detection systems dedicated to the TwinMic beamline at the Elettra synchrotron in Trieste (Italy) and to the XAFS beamline at the SESAME synchrotron in Allan (Jordan) (Fig. 3.3).

The 32-channel low-energy X-ray fluorescence (LEXRF) detector for the Twin-Mic system, a first of its kind to integrate a microscopic imaging and spectroscopic capability, uses a high-gain version of the SIRIO preamplifier, to achieve the best results on low-energy photons. The complete detector has a noticeable total area of $12\,300\,\text{mm}^2$, and thanks to a multi-element SDD architecture allows a wide solid angle coverage of 1.57 steradiants around the specimen, that largely increases the detection efficiency with respect to the previously adopted detection system, and measured a best energy resolution of 125 eV FWHM on the 5.9 keV line [8]. The X-ray absorption and fine structure (XAFS) detector for SESAME, on the other hand, adopts SIRIO to realize a high-count-rate detection system with 64 channels; the detector is capable of reaching 15.5 Mcps, a significant improvement with respect

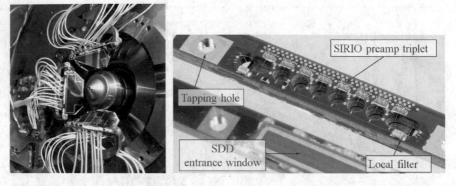

Fig. 3.3 On the left, the TwinMic LEXRF detector. On the right, a detail of the XAFS detector, showing the mounting of the SIRIO preamplifiers on the front-end board. The SDDs are designed by Italian National Institute of Nuclear Physics (INFN) and realized by Fondazione Bruno Kessler (FBK)

to commercially available systems for similar applications, and the best performing pixels registered a resolution below 150 eV FWHM on the 5.9 keV at moderate cooling [9]. The achieved results on the TwinMic/XRF and on the SESAME/XAFS detectors realized by the ReDSoX collaboration, have shown that the SIRIO preamplifier is sufficiently mature to be integrated in complex multi-channel applications. On the other hand, the development of new XAFS detection system at the Elettra synchrotron, based on the one realized for SESAME, but with the possibility of in-vacuum operation, is scheduled in the next few years, which will most likely encourage the realization of a low-power version of the SIRIO-6 preamplifier.

3.3 Application Specific Integrated Circuits for Satellite Instrumentation

The observation of the high energy transients coming from the deep space is a fundamental tool for the understanding of a multitude of astrophysical phenomena. Especially since 2017—when for the first time the simultaneous detection of gravitational waves (GWs) and short gamma ray bursts (GRBs) generated by the same binary neutron star merger has been observed—a new era of multi-messenger astrophysics started: observing the sky with a wide field of view and covering a large frequency range can be extremely useful to trigger the observation and accurate localization of GRBs, which, in turn, enable the search of GWs associated with these high-energy electromagnetic emissions. However, due to the absorption of X and γ photons in the atmosphere, satellite instrumentation is needed for such specific purpose.

Within the THESEUS (Transient High Energy Sky and Early Universe Surveyor) space mission concept, the design of a low-power low-noise ASIC for the readout of the on-board X-γ imaging spectrometer (XGIS) has been carried out. The ASIC, organized in a chipset constellation named ORION, features a dual dynamic range for separate X-ray (2−20 keV) and γ-ray (20–20 000 keV) processing based on a double detection mechanism which combines the usage of SDDs and Thallium activated Cesium-Iodide scintillators. The small power budget, typical of battery operated systems, moved the ASIC toward ultra-low current solutions. On the other hand, the necessity of the XGIS instrument to detect and process with high resolution a wide energy range of incoming photons led to the implementation of a smart on-chip discrimination between X and γ events, which are conveniently processed by two independent channels of the ORION ASIC.

3.3.1 The X and γ Imaging Spectrometer for the THESEUS Space Mission Concept

The on-board XGIS is composed by two cameras, each one based on a large ($\sim$50×50 cm^2) highly segmented detection plane of 6 400 CsI(Tl) crystal scintillator bars ($4.5 \times 4.5 \times 30$ mm^3) optically coupled to 12 800 SDD pixels, realizing a wide field deep sky monitor in the 2 keV–20 MeV energy band [10]. The single detection element, shown in Fig. 3.4, is composed by a scintillator crystal, optically isolated with respect to its neighbours by means of a reflective wrapping, and two SDDs, one at the top and one at the bottom of the bar. The reconstruction of both low energy (<20 keV) and high energy (up to 20 MeV) photons information (energy, position and timing) is possible through appropriate processing of the charge generated by coincident illumination of the top and bottom SDDs.

Among the main challenges faced in the presented ASIC, it must be mentioned the large number of pixels, of relatively large area, present in the concept of the XGIS-camera (12 800 pixels of 25 mm^2 each for the complete system), which oriented the design in a *chipset*, or chip constellation, architecture where a large number of front-end ASICs, placed a few millimeters from the detectors, send a partially shaped signal to a limited number of multi-channel back-end ASICs, which are responsible for the final processing and digitization of the front-end signals. With an on-chip photon discrimination, filtering, analog-to-digital conversion and time-stamping capability, the ORION-BE, realized in collaboration with University of Pavia, can interact directly with an on-board FPGA via SPI protocol.

3.3.2 The ORION Circuit Architecture

A schematic representation of the pixel readout strategy with the ORION chipset is shown in Fig. 3.5, and includes 2×ORION-FE chips and 1×ORION-BE channel. The charge generated by each pixel on the top SDD (in case of a low-energy X event) or

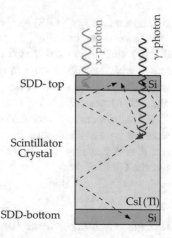

Fig. 3.4 The XGIS double detection operating principle, composed by on CsI(Tl) bar and two SDDs. In such configuration, the radiation entrance window is facing the top SDD of each CsI(Tl) crystal, so that low-energy photons (<20 keV) are absorbed by the SDD at the top interface, while high-energy photons (up to 20 MeV) pass through the top SDD and are absorbed in the CsI(Tl) crystal whose scintillation light illuminates both the SDDs

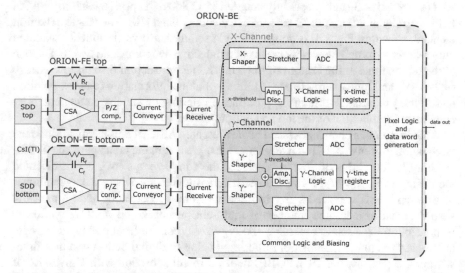

Fig. 3.5 Block-diagram of a single XGIS element readout circuit, composed by two ORION-FE (for the top and bottom SDD pixels) and one ORION-BE with separate X and γ processing channels

by both top and bottom SDDs (in case of a γ event) is processed independently by the ORION-FEs, glued in close proximity of the SDD anodes. The ORION-FE provides the fundamental amplification and processing to the charge signal, and integrates a charge sensitive amplifier in continuous reset configuration—required by timing applications as in the case of THESEUS −, a pole-zero compensation network, and a current conveyor, which is responsible for the first CR shaping and for signal transmission to the ORION-BE. The signals are transmitted in current mode, in an approach already successfully tested on the LYRA chipset, on board the HERMES nanosatellites, which minimizes the cross-talk in multi-channel architectures [11]. The current outputs of the ORION-FE top and bottom channels of the same pixel are delivered to two current receivers in the ORION-BE channel, positioned up to ∼4–5 cm far from the respective FE. Internally, the ORION-BE includes an X-channel, which takes only the signal coming from the top FE, and a γ-channel, which processes the signals coming from both the top and bottom channels.

The X-channel shaping amplifier is a first-order semi Gaussian-Shaper. The shaped signal is compared by an amplitude discriminator with an internally programmable X-threshold; when the amplitude discriminator toggles the peak stretcher is enabled and a peak discrimination block (not shown) recognizes that the peak amplitude of the signal has been reached. The peak detector thus flags to the logic to disconnect the input line, to reduce the probability of pile-up events, and assigns to the photon a unique time-stamp. The ASIC waits a programmable time interval before registering the trigger status, in order to give sufficient time to other channels to settle to a stable value. The stretched output voltage is then over-sampled by a second order 12 bit incremental A/D converter, designed by University of Pavia. When the acquisition at the BE is completed, the central logic sends an end-of-conversion flag, to properly reset the channel for a new acquisition.

At the same time, the signal from the top current receiver is processed also by the γ-channel. In this case, the γ-shaper realizes a third order semi-Gaussian shaping, which is more suitable for the long scintillation time-constants of CsI(Tl) bars, and is replicated also on the bottom channel. The outputs on the top and bottom channels, are summed in current mode to eventually trigger the amplitude discriminator and start the digital processing by the γ channel logic, that operates similarly to the X-channel one. If the γ amplitude discriminator toggles, meaning that a high-energy signal has been detected, the X-channel output is discarded. Two independent stretchers and ADCs are then used to digitize the outputs of the γ shapers, which are eventually summed in the digital domain for a more accurate reconstruction, until an end-of-conversion signal is sent back to reset the analog BE.

3.3.3 Experimental Results

The performances and functionalities of the ORION ASIC coupled to a silicon drift detector have been tested and extensively characterized using a dedicated board designed by INAF-IASF (Milan), shown in Fig. 3.6. The spectroscopic resolution of

Fig. 3.6 First prototype of the test-board used for the ORION-ASIC experimental measurements. The board includes two ORION-FE, one ORION-FE and one multi-element SDD with 4 pixels of 25 mm² each, designed by INFN and produced by Fondazione Bruno Kessler (FBK). The spectra have been acquired sampling the output waveform of the ORION pulse-shaper with a commercial multi-channel analyzer by Amptek

the prototype was tested using ^{241}Am and ^{55}Fe calibration sources to acquire different energy spectra across the expected operating temperature range ($+20°/-20°$ C). The optimum ENC at $-20\,°C$ on the X-channel is 24.3 el. r.m.s. (corresponding to 212 eV FWHM on Si), and 39.6 el. r.m.s. on the γ-channel (corresponding to 3.7 keV FWHM on CsI(Tl)). At the nominal operating temperature of the XGIS module ($+10\,°C$) an ENC of 27.8 el. r.m.s. (241 eV FWHM on Si) on the X-channel, and 43.6 el. r.m.s. (4.1 keV FWHM on CsI(Tl)) γ-channel are recorded.

At the nominal operating temperature of $+10\,°C$, the measured energy threshold of the internal pulse-amplitude discriminator is 563 eV, for X-ray events, and 42 keV for γ-ray events. The analog-processor is able to correctly shape and stretch the input pulses with the designed peaking times (1 μs for X-channel, 3 μs for γ-channel) and producing a look-at-me signal for the ADC when the peak level is reached with a delay between 0.7 and 1 μs. The linearity error of the ASIC in the operative energy range has been measured to be $\pm1\%$ for the X-channel (over an input charge range of $\sim$10 000 el.) and $\pm1.5\%$ for the γ-channel (over an input charge range of $\sim$100 000 el.), a performance which will guarantee an accurate reconstruction of the acquired spectra over the complete 4 decades energy band.

3.4 Concluding Remarks

It is a common misconception to consider nuclear microelectronics only as a supportive and subsidiary mean to the advancement of radiation instrumentation; on the contrary, it should be unmistakable that even mature detection technologies can still reap the fruits of recent progress in microelectronics research. The development of integrated circuits for semiconductor radiation detectors described within this work highlights such an important remark, showing in several applications, how a dedicated electronics can help reaching the ultimate performance limits of silicon drift chambers, set-up new challenges for the room temperature detection, and, eventually, allow the integration of complex processing architectures on-board satellite instrumentation, for the advancement of astrophysics and space science.

Acknowledgements This work was carried out within the ReDSoX project of INFN and the THESEUS project of INAF-ASI. The contribution of universities and research institutes taking part in the collaboration is acknowledged: Polytechnic of Milan, University of Pavia, University of Trieste, University of Udine, University of Bologna, Fondazione Bruno Kessler, Elettra Sincrotrone di Trieste, Italian Space Agency (ASI), Italian National Institute of Astrophysics (INAF), Trento Institute for Fundamental Physics and Applications and Karlsruher Institut für Technologie.

References

1. C. Cottini, E. Gatti, G. Giannelli, G. Rozzi, Minimum noise pre-amplifier for fast ionization chambers. Il Nuovo Cimento (1955-1965) **3**(2), 473–483 (1956)
2. E. Gatti, P. Rehak, Semiconductor drift chamber—an application of a novel charge transport scheme. Nucl Instrum Meth A (1984). https://doi.org/10.1016/0167-5087(84)90113-3
3. G. Gramegna, P. O'connor, P. Rehak, S. Hart, CMOS preamplifier for low-capacitance detectors. Nucl. Instrum. Meth. A (1997). https://doi.org/10.1016/S0168-9002(97)00390-2
4. G. Bertuccio, S. Caccia, Noise minimization of MOSFET input charge amplifiers based on $\Delta\mu$ and ΔN 1/f models. IEEE T Nucl Sci (2009). https://doi.org/10.1109/TNS.2008.2012347
5. F. Mele, J. Quercia, G. Bertuccio, Analytical model of the discharge transient in pulsed-reset charge sensitive amplifiers. IEEE T. Nucl. Sci. (2021). https://doi.org/10.1109/TNS.2021.3087420
6. F. Mele, M. Gandola, G. Bertuccio, SIRIO: a high-speed CMOS charge-sensitive amplifier for high-energy-resolution X-γ ray spectroscopy with semiconductor detectors. IEEE T. Nucl. Sci. (2021). https://doi.org/10.1109/TNS.2021.3055934
7. M. Sammartini, M. Gandola, F. Mele et al., A CdTe pixel detector-CMOS preamplifier for room temperature high sensitivity and energy resolution X and γ ray spectroscopic imaging. Nucl. Instrum. Meth. A (2018). https://doi.org/10.1016/j.nima.2018.09.025
8. J. Bufon, S. Schillani, M. Altissimo et al., A new large solid angle multi-element silicon drift detector system for low energy X-ray fluorescence spectroscopy. J. Instrum. (2018). https://doi.org/10.1088/1748-0221/13/03/C03032
9. A. Rachevski, M. Ahangarianabhari, G. Aquilanti et al., The XAFS fluorescence detector system based on 64 silicon drift detectors for the SESAME synchrotron light source. Nucl. Instrum. Meth. A (2019). https://doi.org/10.1016/j.nima.2018.09.130
10. C. Labanti et al., The x/gamma-ray imaging spectrometer (XGIS) on-board THESEUS: design, main characteristics, and concept of operation, in *Space Telescopes and Instrumentation 2020: Ultraviolet to Gamma Ray* (2020). https://doi.org/10.1117/12.2561012
11. M. Gandola, M. Grassi, F. Mele et al., LYRA: a multi-chip ASIC designed for HERMES X and gamma ray detector, in *IEEE Nuclear Science Symposium and Medical Imaging Conference (NSS/MIC)* (2019). https://doi.org/10.1109/NSS/MIC42101.2019.9059616

Modeling of GIDL–Assisted Erase in 3–D NAND Flash Memory Arrays and Its Employment in NOR Flash–Based Spiking Neural Networks

Gerardo Malavena

Abstract Since the very first introduction of three-dimensional (3–D) vertical-channel (VC) NAND Flash memory arrays, gate-induced drain leakage (GIDL) current has been suggested as a solution to increase the string channel potential to trigger the erase operation. Thanks to that erase scheme, the memory array can be built directly on the top of a n^+ plate, without requiring any p-doped region to contact the string channel and therefore allowing to simplify the manufacturing process and increase the array integration density. For those reasons, the understanding of the physical phenomena occurring in the string when GIDL is triggered is important for the proper design of the cell structure and of the voltage waveforms adopted during erase. Even though a detailed comprehension of the GIDL phenomenology can be achieved by means of technology computer-aided design (TCAD) simulations, they are usually time and resource consuming, especially when realistic string structures with many word-lines (WLs) are considered. In this chapter, an analysis of the GIDL-assisted erase in 3–D VC NAND memory arrays is presented. First, the evolution of the string potential and GIDL current during erase is investigated by means of TCAD simulations; then, a compact model able to reproduce both the string dynamics and the threshold voltage transients with reduced computational effort is presented. The developed compact model is proven to be a valuable tool for the optimization of the array performance during erase assisted by GIDL. Then, the idea of taking advantage of GIDL for the erase operation is exported to the context of spiking neural networks (SNNs) based on NOR Flash memory arrays, which require operational schemes that allow single-cell selectivity during both cell program and cell erase. To overcome the block erase typical of NOR Flash memory arrays based on Fowler-Nordheim tunneling, a new erase scheme that triggers GIDL in the NOR Flash cell and exploits hot-hole injection (HHI) at its drain side to accomplish the erase operation is presented. Using that scheme, spike-timing dependent plasticity (STDP) is implemented in a mainstream NOR Flash array and array learning is successfully demonstrated in a prototype SNN. The achieved results represent an important step for the development of large-scale neuromorphic systems based on mature and reliable memory technologies.

G. Malavena (✉)
Dipartimento di Elettronica, Informazione e Bioingegneria, Politecnico di Milano, Via Ponzio 34/5, 20133 Milano, Italy
e-mail: gerardo.malavena@polimi.it

L. Piroddi (ed.), *Special Topics in Information Technology*,
PoliMI SpringerBriefs, https://doi.org/10.1007/978-3-030-85918-3_4

Keywords 3–D NAND Flash memory arrays · NOR Flash memory arrays · GIDL · Neuromorphic computing · Spiking Neural Networks · Artificial Neural Networks

1 Introduction

Since their very first introduction, the performance improvement of Flash memory technologies was long achieved thanks to an uninterrupted scaling process that led to a NAND Flash cell feature size as small as 14 nm in 2015 [1]. However, as the size of the single memory cell was shrinked down to decananometer dimensions, some fundamental issues related to the increasingly complex fabrication techniques and to inherent physical limitations due to the discrete nature of charge and matter emerged, undermining both the manufacturing and the proper operation of Flash memory arrays [2, 3]. For this reason, an alternative integration paradigm has been adopted to break the classical trade-off between single-cell area and array storage density, consisting in stacking many layers of memory cells along the direction orthogonal to the wafer plane. Figure 1a, shows a schematic of a possible implementation of 3-D NAND Flash memory arrays, featuring vertical polycrystalline silicon (poly-Si) channels, contacted at the bit-line (BL) and source-line (SL) sides by n^+ regions. At the intersection between each poly-Si channel and a horizontal word-line (WL) plane, a gate-all-around (GAA) memory cell with *macaroni* structure is formed, as schematically depicted in Fig. 1b, with the oxide-nitride-oxide (ONO) stack adopted to store charge in the middle layer. Due to the lack of any p-doped regions to access the string channel in such architecture, the poly-Si channels cannot be contacted similarly to the case of planar NAND technologies. While this feature does not affect the read and program operations, the employment of a novel voltage scheme is required to increase the channel potential and trigger the emission of electrons from the storage layer or the injection of holes into it during erase. To this purpose, the voltage scheme displayed in Fig. 1c is adopted. A positive voltage ramp is applied to the BL and SL of the string while keeping to ground the WLs and the selector gates (SGs); the strong electric fields at the inner edge of the SGs are large enough to trigger the generation of electron/hole pairs by band-to-band tunneling (BTBT) [4, 5]. While electrons are swept towards the BL/SL contacts, giving rise to the so-called GIDL current, the BTBT-generated holes are directed towards the center of the string, where they accumulate and contribute to increase the channel potential. In this framework, Sect. 2 is devoted to the study of the GIDL-assisted erase in 3–D NAND Flash memory arrays by TCAD simulations first (Sect. 2.1), and, then, to the development of a compact model able to predict the string behavior and the threshold-voltage V_T evolution during erase (Sect. 2.2). All the presented results are from [6] and [7].

On the other hand, NOR Flash memory cells have never been scaled beyond the feature size of 40 nm as research efforts for embedded applications have been focused on different technologies, such as phase-change memories [9, 10]. Despite this, in the last years NOR Flash memory arrays attracted some interest also for their

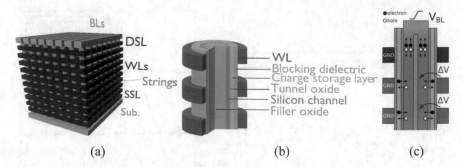

Fig. 1 **a** Schematic of a VC 3–D NAND Flash memory array and **b** of a GAA memory cell [1] (from [8] under CC BY 4.0 license). **c** Schematic of the string condition during a GIDL-assisted erase operation (a section of the string close to the BL is shown)

employment in the implementation of spiking neural networks (SNNs), representing a promising solution to outclass conventional CMOS systems based on the Von-Neumann architecture in problems dealing with unstructured data, such as image recognition and classification [11]. A mandatory condition to be met by memory arrays employed in SNNs is the possibility to tune the threshold voltage (V_T) of each cell independently of the others in both directions, meaning that single-cell selectivity not only during program and but also during erase operation is needed, with the block erase typical of Flash technologies clearly representing an obstacle for neuromorphic applications. To overcome this issue, some works have suggested design adjustments either to the cell or to the array level, with the drawback, however, of a larger array area occupancy and more complex manufacturing process [12–15]. Taking inspiration from the GIDL-assisted erase employed in 3–D VC NAND Flash memory arrays investigated in Sect. 2, the idea of moving from the classical erase scheme based on Fowler-Nordheim (FN) tunneling [16] to a novel single-cell selective one that exploits BTBT-generated HHI at the drain side is presented in Sect. 3. Exploiting such scheme, the operation of a SNN based on the STDP learning rule [17–19] exploiting a mainstream NOR Flash memory array with no modification either to the cell or to the array design is successfully demonstrated. The results presented in Sect. 3 are from [20–23].

2 GIDL–Assisted Erase in 3-D NAND Memory Arrays

2.1 Overview on String Dynamics

In order to investigate the erase operation in 3-D NAND Flash memory arrays when GIDL is triggered at the SGs, TCAD simulations were performed using a commercial device simulator (see [6] for more information about the simulation environment). As a starting point, no charge exchange between the channel and the storage layer

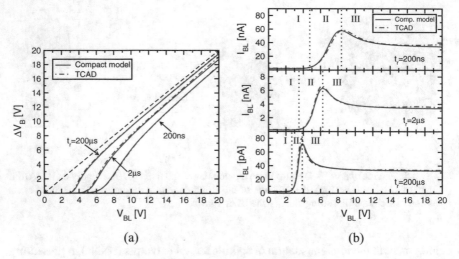

(a)   (b)

Fig. 2 **a** Comparison between the string potential V_B during the erase transient resulting from the TCAD simulations and that computed with the developed compact model. **b** Same as **a** but I_{BL} is shown. © 2018 IEEE [6]

is included. Simulation results are displayed in Fig. 2a and Fig. 2b, which report the variations of the string potential ΔV_B (the average value in the radial direction at the center of the string is considered) and the BL/SL currents $I_{BL} = I_{SL}$ during erase for different values of the BL/SL ramps rise time t_r. Results reveal that three different phases of the transient can be identified: I) $I_{BL} \approx$ constant and $\Delta V_B \approx 0$; II) I_{BL} increases steeply and the same does V_B, with rate larger than that of V_{BL}; III) I_{BL} reaches a peak and then saturates to a constant value while V_B continues to increase but at the same rate of V_{BL}. Figure 3 shows how the net charge density in the NAND string evolves during the erase transient. By comparing the former figure with Fig. 2a and Fig. 2b, it is easy to relate the charge distribution with the ΔV_B and I_{BL} transients: during phase I, BTBT-generated holes are a few and the string is approximately depleted of charge; during phase II, holes start to rule the string electrostatics, but they are confined in the central part of the string (that is, the SGs regions are still depleted); during phase III BTBT-generated holes spread also under the SGs, thus ruling their electrostatics.

2.2 Compact Model

Figure 4a shows the compact model developed to reproduce the results of the TCAD simulations. Holes distribution in the string is approximated to be uniform (orange region) over an equipotential region that extends from the center of the string to a distance ΔL within the channel of the SG, which is variable during the transient.

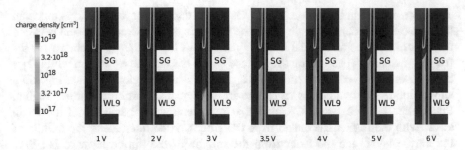

Fig. 3 TCAD results displaying the net charge density (normalized to the elementary charge) at different times during the same transients of Fig. 2a and Fig. 2b. Below each snapshot the V_{BL} at which it is taken is reported. Note that the net charge density in the central region of the string is due to holes, while that at the bottom of the n^+-doped regions is due to ionized donors. © 2018 IEEE [6]

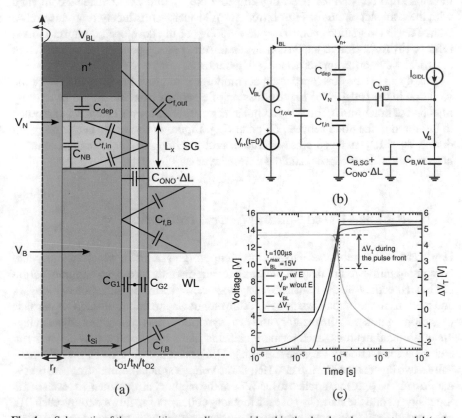

Fig. 4 **a** Schematic of the capacitive couplings considered in the developed compact model (only the upper region of the string close to the BL is showed, © 2019 Springer Nature [7]). **b** Schematic of the developed compact model. **c** Evolution of V_B and ΔV_T (with respect to the neutral V_T) during erase when also the charge exchange between the string channel and the nitride storage layer is included in the compact model of Fig. 4b. (© 2019 Springer Nature [7])

The electrostatics in the region of the SGs that is depleted of holes (L_x) is modeled through $C_{f,in}$ and C_{NB}; $C_{ONO} \cdot \Delta L$ is the remaining capacitive component between the bulk orange region and the longitudinal face of the SG; $C_{f,out}$ accounts for the fringing fields between the transverse face of the SG and the n^+ region while $C_{f,B}$ between the transverse face of the WLs and the central region of the string; C_{dep} simply models the variations of charge in the depleted portion of the n^+ region. Finally, the series between C_{G1} and C_{G2} represents the capacitance of the ONO stack, with the former calculated from the silicon/oxide interface to the middle of the nitride layer, and the latter from the middle of the nitride layer to the WL. The resulting compact model is shown in Fig. 4b, with the addition of the current generator I_{GIDL} that reproduces the GIDL current ($C_{B,SG}$ and $C_{B,WL}$ are the overall capacitances between the orange region and the SGs and WLs, respectively). Please refer to [6] and [7] for the calculation of I_{GIDL} and of all the capacitive contributions mentioned so far. Figure 2a and Fig. 2b show the V_B and I_{BL} transients computed with the compact circuit of Fig. 4b (red line). Model results nicely reproduce those from TCAD simulations confirming the validity of the developed compact model; refer to [6] for a similar analysis also for different string geometries and different electrical waveforms applied to the string contacts.

Finally, in [7] the developed compact model was improved to account also for the variation in the cell V_T due to the emission of electrons from the nitride layer or by injection of holes into it; for compact modeling purposes, the net charge was assumed to be stored in the node between C_{G1} and C_{G2}. Figure 4c shows the evolution of V_T during the GIDL-assisted erase operation and the impact of the charge exchange between the silicon channel and the nitride layer on V_B.

3 NOR Flash–Based Spiking Neural Networks

Hardware neural networks (HNNs) are computing systems in which memory and computing units are not distinct entities exchanging data through a communication bus but rather they are distributed in a way that resembles the organization of synapses and neurons in the human brain [24]. A convenient way to implement HNNs consists in exploiting non-volatile memory arrays as synaptic arrays connecting adjacent layers of artificial neurons: each memory cell acts like a biological synapse, that is, an electrical connection of variable strength [11]. For example, Fig. 5a shows schematically a two layers NOR Flash-based HNN. The voltage signals coming from the presynaptic neurons (PRE) are applied to the WLs of the memory arrays; then, as result of the input signals and the state (V_T) of each memory cell, a current flows through each BL, corresponding to the output signals that are sent to the postsynaptic neurons (POST). In particular, each memory cell is operated in subthreshold regime [25, 26], in which the drain-to-source current I_{DS} can be expressed as a function of the WL voltage V_{WL} as $I_{DS} = I_0 \cdot \exp\left[\alpha_G\left(V_{WL} - V_T^{ref}\right)/(mkT)\right] \cdot \exp\left[\alpha_G \Delta V_T/(mkT)\right]$, where ΔV_T is the cell V_T shift from a reference condition V_T^{ref} (see [23] for the remaining

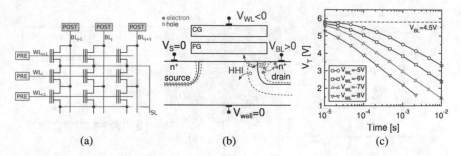

Fig. 5 **a** Schematic of a NOR Flash memory array used as a synaptic arrays connecting two layers of neurons. **b** Schematic a NOR Flash cell subjected to the bias scheme used to trigger HHI and **c** the resulting experimental V_T erase transients (© 2018 IEEE [23])

parameters). In the previous equation the scaling factor $w = \exp[\alpha_G \Delta V_T/(mkT)]$, which is a function of V_T but not of V_{WL}, plays the role of the synaptic weight and the remaining one corresponds to the input signal. A HNN specializes its behavior to perform a well defined task after a *learning* phase, during which the weights of all the memory cells are tuned according to specific *learning algorithms* or *learning rules*. Spiking Neural Networks (SNNs) are particular HNNs for which learning is carried out according to biologically inspired learning rules without external supervision, such as STDP; they take their name from the integrate-and-fire behavior of the artificial neurons, delivering asynchronous spikes during network operation [11, 27].

3.1 Implementing STDP and Unsupervised Learning

Figure 5b shows a schematic of the erase scheme devised to overcome to achieve single-selectivity during erase to enable the adoption of mainstream NOR Flash memory arrays in SNNs. By applying simultaneously a positive V_{BL} and a negative V_{WL}, holes, generated by BTBT and accelerated by the horizontal electric field, become energetic enough to overcome the Si/SiO_2 energy barrier and to be injected into the cell floating-gate FG, leading to $\Delta V_T < 0$. Figure 5c displays the resulting V_T transients, measured for $V_{BL} = 4.5$ V and different values of V_{WL}, confirming the feasibility of the suggested erase scheme.

Once single-cell selectivity during erase is achieved, it is possible to implement STDP in the NOR Flash array, according to which w variations of each memory cell must depend only on the timing between the presynaptic spike and the postsynaptic one (Δt). To that purpose, the voltage scheme displayed in Fig. 6a was devised, exploiting HHI for erase and the classically adopted channel hot-electron injection (CHEI) [28] for program. A presynaptic spike triggers a double-triangular WL pulse of duration t_{WL}; the postsynaptic spike, instead, results in the application of a rectangular pulse to the BL of amplitude equal to 4.5 V, duration much shorter than t_{WL} and delayed with respect to the fire event of $t_{WL}/2$. According to such scheme, if $\Delta t > 0$,

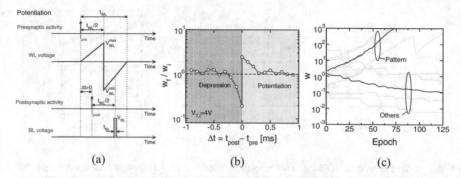

(a)   (b)   (c)

Fig. 6 **a** Voltage scheme suggested to implement STDP in a NOR Flash cell exploiting CHEI and HHI (the $\Delta t > 0$ case is shown) and **b** the resulting experimental STDP waveform. **c** Evolution during the learning phase of the weights of the memory cells in the prototype SNN; the blue curve is the average of the weights belonging to the input pattern, while the red one is the average of the remaining ones. © 2019 IEEE [21]

the BL pulse is applied in correspondence of a negative V_{WL}, thus triggering HHI that results in $\Delta V_T < 0$ ($\Delta w > 0$). In the opposite case, CHEI results in $\Delta V_T > 0$ ($\Delta w < 0$). The experimental STDP waveform resulting from the implementation of the scheme of Fig. 6a is reported in Fig. 6b, displaying that the final weight w_f after a fire event depends on Δt similarly to what observed on biological synapses [11, 17].

Finally, starting from the STDP scheme of Fig. 6a, a prototype SNN with 8 input signals e 1 output was implemented and tested in a pattern learning problem. The input pattern was encoded in the activity of the input neurons, meaning that neurons that are part of the pattern continuously deliver input spikes, otherwise their outputs are kept to ground. The input pattern is correctly *learned* by the SNN if the weights of the cells belonging to it are potentiated and the remaining ones are depressed, as demonstrated in Fig. 6c for the implemented SNN. Please refer to [21, 23] for a full discussion.

Besides, it is worth mentioning that when the employment of NOR Flash memory arrays in HNNs is considered, the impact of their non-idealities during V_T-tuning processes and their typical reliability issues must be carefully assessed. As a matter of example, in [22] the impact of program noise [29] and random telegraph noise [30, 31] on the performance of a neuromorphic digit classifier is investigated in detail. From the suggested analysis, also some quantitative criteria to determine how scaled NOR Flash cells can be when targeting neuromorphic applications are provided.

4   Conclusions

In this chapter, the GIDL-assisted erase operation in 3–D NAND Flash memory arrays has been investigated by means of TCAD simulations and a compact model to reproduce the evolution of I_{BL}, V_B and the cell V_T has been presented. Thanks to its simplicity and accuracy, the model represents a valuable tool for the optimization of the array performance during erase assisted by GIDL. Then, a similar erase scheme has been employed also in NOR Flash memory arrays, exploiting BTBT-generated HHI to enable single-cell selectivity during erase and allowing the adoption of mainstream NOR Flash memory arrays in SNNs without any modification either to the cell or to the array design. The presented results pave the way towards the development of neuromorphic systems based on cost-effective and highly-reliable memory arrays.

References

1. C. Monzio Compagnoni, A. Goda, A. Sottocornola Spinelli, P. Feeley, A.L. Lacaita, A. Visconti, Reviewing the evolution of the NAND flash technology. Proc. IEEE **105**(9), 1609–1633 (2017)
2. C. Monzio Compagnoni, A. Sottocornola Spinelli, Reliability of NAND Flash arrays: a review of what the 2-D-to-3-D transition meant. IEEE Trans. Electron Devices **66**(11), 4504–4516 (2019)
3. A. Sottocornola Spinelli, C. Monzio Compagnoni, A.L. Lacaita, Reliability of NAND Flash memories: planar cells and emerging issues in 3D devices. Computers **6**(2), 16 (2017)
4. T. Chan, J. Chen, P. Ko, C. Hu, The impact of gate-induced drain leakage current on MOSFET scaling, in *1987 International Electron Devices Meeting (IEDM)*. (IEEE, 1987), pp. 718–721
5. J. Fan, M. Li, X. Xu, Y. Yang, H. Xuan, R. Huang, Insight into gate-induced drain leakage in silicon nanowire transistors. IEEE Trans. Electron Devices **62**(1), 213–219 (2014)
6. G. Malavena, A.L. Lacaita, A. Sottocornola Spinelli, C. Monzio Compagnoni, Investigation and compact modeling of the time dynamics of the GIDL-assisted increase of the string potential in 3-D NAND Flash arrays. IEEE Trans. Electron Devices **65**(7), 2804–2811 (2018)
7. G. Malavena, A. Mannara, A.L. Lacaita, A. Sottocornola Spinelli, C. Monzio Compagnoni, Compact modeling of GIDL-assisted erase in 3-D NAND Flash strings. J. Comput. Electron. **18**(2), 561–568 (2019)
8. A. Sottocornola Spinelli, G. Malavena, A.L. Lacaita, C. Monzio Compagnoni, Random telegraph noise in 3-D NAND Flash memories. Micromachines **12**(6), 703–717 (2021)
9. S. Lai, Current status of the phase change memory and its future, in *2003 International Electron Devices Meeting (IEDM)* (IEEE, 2003), pp. 255–258
10. C. Monzio Compagnoni, , Gusmeroli, R., Sottocornola Spinelli, A., Ielmini, D., Lacaita, A.L., Visconti, A.: Present status and scaling challenges for the NOR Flash memory technology (Chap. 3), in *Solid State Electronics Research Advances*, ed. by S. Kobadze (Nova Science Publishers, Inc., 2009), pp. 101–134
11. G.W. Burr, R.M. Shelby, A. Sebastian, S. Kim, S. Kim, S. Sidler, K. Virwani, M. Ishii, P. Narayanan, A. Fumarola et al., Neuromorphic computing using non-volatile memory. Adv. Phys. X **2**(1), 89–124 (2017)
12. F.M. Bayat, X. Guo, H. Om'Mani, N. Do, K.K. Likharev, D.B. Strukov, Redesigning commercial floating-gate memory for analog computing applications, in *2015 IEEE International Symposium on Circuits and Systems*. (IEEE, 2015), pp. 1921–1924
13. C.H. Kim, S. Lee, S.Y. Woo, W.M. Kang, S. Lim, J.H. Bae, J. Kim, J.H. Lee, Demonstration of unsupervised learning with spike-timing-dependent plasticity using a TFT-type NOR Flash memory array. IEEE Trans. Electron Devices **65**(5), 1774–1780 (2018)
14. H. Kim, S. Hwang, J. Park, S. Yun, J.H. Lee, B.G. Park, Spiking neural network using synaptic transistors and neuron circuits for pattern recognition with noisy images. IEEE Electron Device Lett. **39**(4), 630–633 (2018)

15. F. Merrikh-Bayat, X. Guo, M. Klachko, M. Prezioso, K.K. Likharev, D.B. Strukov, High-performance mixed-signal neurocomputing with nanoscale floating-gate memory cell arrays. IEEE Trans. Neural Netw. Learn. Syst. **29**(10), 4782–4790 (2017)
16. M. Lenzlinger, E. Snow, Fowler-Nordheim tunneling into thermally grown SiO_2. J. Appl. Phys. **40**(1), 278–283 (1969)
17. G. Bi, M. Poo, Synaptic modifications in cultured hippocampal neurons: dependence on spike timing, synaptic strength, and postsynaptic cell type. J. Neurosci. **18**(24), 10464–10472 (1998)
18. D.O. Hebb, *The Organization of Behavior: A Neuropsychological Theory* (Wiley, Chapman & Hall, 1949)
19. S. Lowel, W. Singer, Selection of intrinsic horizontal connections in the visual cortex by correlated neuronal activity. Science **255**(5041), 209–212 (1992)
20. G. Malavena, M. Filippi, A. Sottocornola Spinelli, C. Monzio Compagnoni, Unsupervised learning by spike-timing-dependent plasticity in a mainstream NOR Flash memory array-part I: cell operation. IEEE Trans. Electron Devices **66**(11), 4727–4732 (2019)
21. G. Malavena, M. Filippi, A. Sottocornola Spinelli, C. Monzio Compagnoni, Unsupervised learning by spike-timing-dependent plasticity in a mainstream NOR Flash memory array-Part II: array learning. IEEE Trans. Electron Devices **66**(11), 4733–4738 (2019)
22. G. Malavena, S. Petrò, A. Sottocornola Spinelli, C. Monzio Compagnoni, Impact of program accuracy and random telegraph noise on the performance of a NOR Flash-based neuromorphic classifier, in *ESSDERC 2019-49th European Solid-State Device Research Conference (ESSDERC)* (IEEE, 2019), pp. 122–125
23. G. Malavena, A. Sottocornola Spinelli, C. Monzio Compagnoni, Implementing spike-timing-dependent plasticity and unsupervised learning in a mainstream NOR Flash memory array, in *2018 IEEE International Electron Devices Meeting (IEDM)* (IEEE, 2018), pp. 2.3.1–2.3.4
24. P. Sterling, S. Laughlin, *Principles of Neural Design* (MIT Press, s2015)
25. C. Diorio, P. Hasler, B.A. Minch, C.A. Mead, A floating-gate MOS learning array with locally computed weight updates. IEEE Trans. Electron Devices **44**(12), 2281–2289 (1997)
26. P.E. Hasler, C. Diorio, B.A. Minch, C. Mead, Single transistor learning synapses, in *Advances in Neural Information Processing Systems* (1995), pp. 817–824
27. W. Gerstner, W.M. Kistler, R. Naud, L. Paninski, *Neuronal Dynamics: From Single Neurons to Networks and Models of Cognition* (Cambridge University Press, 2014)
28. B. Eitan, D. Frohman-Bentchkowsky, Hot-electron injection into the oxide in n-channel MOS devices. IEEE Trans. Electron Devices **28**(3), 328–340 (1981)
29. C. Monzio Compagnoni, A. Sottocornola Spinelli, R. Gusmeroli, S. Beltrami, A. Ghetti, A. Visconti, Ultimate accuracy for the NAND Flash program algorithm due to the electron injection statistics. IEEE Trans. Electron Devices **55**(10), 2695–2702 (2008)
30. A. Ghetti, C. Monzio Compagnoni, A. Sottocornola Spinelli, A. Visconti, Comprehensive analysis of random telegraph noise instability and its scaling in deca-nanometer Flash memories. IEEE Trans. Electron Devices **56**(8), 1746–1752 (2009)
31. C. Monzio Compagnoni, R. Gusmeroli, A. Spinelli Sottocornola, A.L. Lacaita, M. Bonanomi, A. Visconti, Statistical model for random telegraph noise in Flash memories. IEEE Trans. Electron Devices **55**(1), 388–395 (2007)

Low-Noise Mixed-Signal Electronics for Closed-Loop Control of Complex Photonic Circuits

Francesco Zanetto

Abstract An increasing research effort is being carried out to profit from the advantages of photonics not only in long-range telecommunications but also at short distances, to implement board-to-board or chip-to-chip interconnections. In this context, Silicon Photonics emerged as a promising technology, allowing to integrate optical devices in a small silicon chip. However, the integration density made possible by Silicon Photonics revealed the difficulty of operating complex optical architectures in an open-loop way, due to their high sensitivity to fabrication parameters and temperature variations. In this chapter, a low-noise mixed-signal electronic platform implementing feedback control of complex optical architectures is presented. The system exploits the ContactLess Integrated Photonic Probe, a non-invasive detector that senses light in silicon waveguides by measuring their electrical conductance. The CLIPP readout resolution has been maximized thanks to the design of a low-noise multichannel ASIC, achieving an accuracy better than $-35\,\mathrm{dBm}$ in light monitoring. The feedback loop to stabilize the behaviour of photonic circuits is then closed in the digital domain by a custom mixed-signal electronic platform. Experimental demonstrations of optical communications at high data-rate confirm the effectiveness of the proposed approach.

1 Introduction

The increasing demands in terms of bandwidth and energy efficiency required by telecommunications, automotive applications and datacenters are pushing copper-based interconnections close to their intrinsic limits [1]. In the same way, the new trends on rack-scale multi-processor computational units are putting pressure on chip-to-chip connections, calling for low-latency high-speed performance at reduced power and cost [2]. These technological requirements are making traditional interconnects the bottleneck of high-performance systems, suggesting the use of point-

F. Zanetto (✉)
Dipartimento di Elettronica, Informazione e Bioingegneria, Politecnico di Milano, Via Ponzio 34/5, 20133 Milano, Italy
e-mail: francesco.zanetto@polimi.it

© The Author(s) 2022
L. Piroddi (ed.), *Special Topics in Information Technology*,
PoliMI SpringerBriefs, https://doi.org/10.1007/978-3-030-85918-3_5

to-point optical connectivity not only for long distance communications but also at short range [3]. Compared to other solid-state solutions, Silicon Photonics (SiP) seems to be the ideal candidate to answer to these needs, sharing the same fabrication technology of the microelectronic industry.

Similarly to MEMS technologies, which exploited the mechanical properties of silicon to create a new family of devices and applications that are now supporting the electronic industry with a large fraction of its revenues [4, 5], Silicon Photonics is called to leverage the optical properties of silicon to create a new realm of minia-turized systems and devices and implement innovative functionalities. The silicon crystal is in fact transparent to near-infrared radiation, in particular to 1550 nm, which is the typical working wavelength of long-range fiber links, and to 1310 nm, which is of increasing interest for short-distance interconnects. Moreover, light confinement into a lithographed waveguide can be obtained by using silicon dioxide as cladding material, thanks to the large refractive index difference with respect to silicon. These features have already enabled the demonstration of high-complexity architectures integrating different kind of photonic devices, like Mach Zehnder interferometers (MZI), ring resonators, arrayed waveguide grating router (AWGR) [6, 7], allowing on-chip manipulation of light.

Even though photonic technologies have already demonstrated maturity for inte-grating lots of devices in a small footprint [8], the widespread use of complex silicon photonic architectures is still limited in real applications. The motivation for this delay is found in the inherent nature of photonic devices, that usually rely on inter-ferometry and are thus very sensitive to fabrication tolerances, temperature variations and mutual crosstalk. As an example, the resonant frequency of a ring resonator filter is observed to move of 10 GHz due to 1 °C temperature change [9], making it very hard to reliably operate photonic architectures in harsh environments like datacen-ters without a real-time monitoring of their working conditions. Local light sensing, possibly with CMOS compatible detectors, and active control of each photonic ele-ment thus emerged as strong requirements to implement closed-loop stabilization and fully exploit the potential of photonic integration.

2   The Challenge of Transparent Detection

To reliably operate a complex photonic circuit, the working point of each integrated device needs to be assessed in real-time without changing its functionality. When the number of elements in a circuit is limited to a few units, light monitoring can be obtained by tapping a small part of the optical power from the waveguide towards a photodetector. The use of germanium photodiodes is the most common solution in this case [10, 11], even though the addition of this material to the technology is not trivial and it increases the final cost of the die. However, when the count of inte-grated devices in the chip increases to hundreds or thousands, this approach becomes rapidly unfeasible, as the large quantity of probing points causes an unacceptable light attenuation and/or perturbation [12].

For these reasons, a relevant research effort is being carried out to investigate the use of non-invasive in-line photodetectors, that exploit the waveguide itself as a light sensor and promise to allow the control of large-scale photonic circuits. Even though silicon is nominally transparent to near-infrared wavelengths, photocarrier generation has been observed in silicon waveguides because of two photon absorption (TPA) [13, 14] and sub-bandgap mechanisms such as surface-state absorption (SSA) [15, 16], that are responsible of the intrinsic propagation losses of photonic waveguides. Since the amount of free carriers generated by these mechanisms is small, transparent detectors usually have a lower sensitivity with respect to germanium photodiodes but, as they detect the full optical power in the waveguide and not just a small part of it, their use is not so disadvantageous compared to the standard approach.

Among the several solutions found in literature, the fully transparent ContactLess Integrated Photonic Probe (CLIPP) [17] was chosen in this work to monitor the state of complex photonic circuits. Like a photoconductor, this detector measures the changes of waveguide conductance caused by free carriers generation. However, to perform a truly non-invasive sensing, the core is not doped or processed and the access to its electrical properties is obtained capacitively, by adding two metal electrodes on top of the cladding 700 nm away from the core. Figure 1a shows a 3D picture of the CLIPP detector and its simplified electrical model. To effectively bypass the access capacitance C_A and read the electrical resistance of the waveguide R_{WG}, one electrode of the sensor is driven at a frequency $f_{CLIPP} > 1/\pi C_A R_{WG}$, usually between 100 kHz and 10 MHz depending on the pad geometry, while the other is connected to a TransImpedance Amplifier (TIA) to collect the current of the sensor and translate it into a voltage. A lock-in detection, achieved by demodulating the output of the TIA at the same frequency of the sensor stimulation, is thus a convenient readout scheme, allowing to reconstruct the complex admittance of the sensor and minimize the electronic noise at the same time. The readout noise is indeed of particular importance in this application. The conductance of the core is around 1 nS [17], so a resolution better than 1 pS is targeted to monitor the optical power in photonic devices with an accuracy below −30 dBm (Fig. 1b).

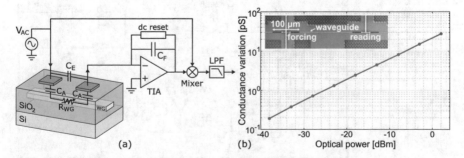

(a) (b)

Fig. 1 **a** 3D representation of a CLIPP detector and its readout. **b** Sensitivity curve of a 400 μm device. Low noise readout is required to detect light variations below −30 dBm, corresponding to conductance changes <1 pS

3 Integrated Lock-In Readout System

To perform an effective CLIPP readout with the best possible resolution, a multichannel lock-in based ASIC was specifically designed in STMicroelectronics BCD8sP 0.18 μm technology (Fig. 2) [18]. The design takes advantage of a capacitive-feedback TIA architecture to simultaneously achieve low noise and wide bandwidth. The front-end amplifier features a series white noise of only $2.8 \, \text{nV}/\sqrt{\text{Hz}}$, with a corner frequency of 300 kHz, and a gain-bandwidth product of 2 GHz, reflecting in a closed-loop bandwidth around 80 MHz. The capacitive feedback, implemented with an inherently linear passive element, guarantees the required linearity over a rail-to-rail output.

Capacitive-feedback TIAs require a DC handling network to set the bias of the amplifier and discharge any leakage current coming from the sensor, that would otherwise cause the saturation of the output. This problem is of particular relevance in photonic applications, where the electronic front-end is used close to the photonic devices and light escaping from the optical circuit can cause photo-generated currents at the input of the TIA up to the nA range. To solve this issue, an active DC feedback network has been implemented. Here, the output voltage of the TIA is read by an auxiliary active circuit, which removes the AC signal and feeds the DC and low-frequency components back to the input node, effectively creating a DC handling mechanism. By using an integrator in the feedback path, this topology operates with a zero residual offset, apart from mismatch effects and non-idealities.

The overall gain of the readout circuit, which depends on the value of the feedback capacitance C_F and on the amplitude of the signal used to stimulate the CLIPP sensor, is ultimately limited by the value of the feedthrough capacitance C_E. To mitigate this issue and increase the maximum stimulation amplitude, a capacitance compensation system was added at the input of the chip. A programmable capacitor, tuned with a 4-bit digital-to-capacitance converter (DCC) and driven in counter-phase with respect to the stimulation of the sensor, is connected in parallel to the CLIPP to sink the

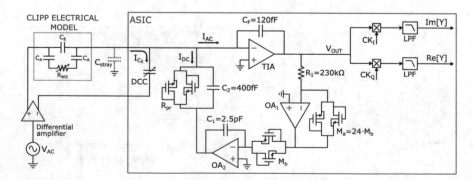

Fig. 2 Schematic of the 11-channel low-noise wide-bandwidth lock-in front-end ASIC, fabricated in 0.18 μm STMicroelectronics BCD8sP technology

spurious current injected by C_E. This solution allows to use stimulation signals as high as 10 V with a C_F of 120 fF, with great benefits in terms of readout resolution. Indeed the DCC, together with the input-referred current noise below $1 \, pA/\sqrt{Hz}$, allows to achieve a resolution in the readout of conductance well below 1 pS. This extreme level of accuracy enables the detection of light variations below $-35 \, dBm$, suitable to control photonic architectures with the required precision.

4   Multichannel Electronic Platform for Closed-Loop Control of Photonic Circuits

The analog preamplification performed by the integrated front-end is complemented by an electronic system to reliably control the working point of photonic circuits. The system is designed in a modular way (Fig. 3) [19]: a first PCB, shaped to best fit in the optical bench, houses the photonic and electronic chips, while a larger motherboard hosts the rest of the electronics. This approach allows to maximize the readout accuracy by mounting the two chips close to each other, while allowing easy optical coupling and good flexibility in the design of the electronics.

The custom mixed-signal motherboard is designed to: (i) generate the sinusoidal signal necessary to drive the CLIPP, with tunable frequency (50 kHz–10 MHz) and amplitude (1–10 V) to adapt to different sensor geometries; (ii) amplify, filter and D/A convert the signal coming from 16 CLIPP detectors, without introducing any degradation in the readout resolution; (iii) drive up to 16 actuators on the photonic chip, with a voltage precision of few mV and maximum current of 50 mA; (iv) perform the feedback algorithms to control the working points of photonic devices.

The digital core of the system is an FPGA for real-time flexible parallel processing. The FPGA is housed in a commercial module, with all the components to connect it with a PC via USB. The FPGA handles the communication with all the components of the platform and generates the signals needed for the lock-in detection. To achieve the best possible resolution in the measurement, a heterodyne lock-in down-conversion is performed: a first analog mixer in the ASIC moves the signal to an intermediate frequency (f_{MID}, around some kHz) above the 1/f noise corner frequency of the acquisition chain, then, after digitization with the ADC, a second digital I/Q demodulation is done in the FPGA to down-convert the signal to baseband as required by the lock-in processing. A tunable digital filter sets the final readout bandwidth, which is defined based on the accuracy and speed requirements of the application.

The FPGA also implements the strategies to control the photonic devices. The operations can be divided into tuning, i.e. scanning the heaters voltage until the maximum or minimum of a desired cost function is reached, or locking, i.e. real-time feedback stabilization of the working point of a photonic device. Different control laws can be implemented and we successfully demonstrated integral and PI controllers [20], as well as gradient methods. The FPGA also generates modu-

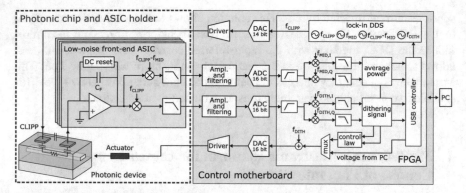

Fig. 3 Schematic view of the mixed-signal control platform for automated control of photonic circuits

lation tones (indicated as f_{DITH} in Fig. 3), which can be used for two reasons: (i) "labelling" each wavelength by modulating the input light source, to discriminate different signals in the optical architecture; (ii) using the dithering technique, by adding the tones to the heaters voltage to obtain a signal proportional to transfer function derivative of a photonic device, useful for power-independent locking [21]. A bank of additional lock-in mixers and filters is implemented in the digital domain for an efficient detection of these tones.

5 Experimental Demonstrations

An experimental demonstration of the full electronic system has been carried out with a photonic architecture conceived for intra-datacenter optical communications. The chip implements a low-energy any-to-any light router between N processing units, a very promising approach since it solves the radix-latency of switch-based solutions normally used to connect sockets inside datacenters. Each processor is equipped with a transmission engine (Tx), comprising N-1 lasers at different wavelengths that are combined on a single waveguide by an optical (N-1):1 multiplexer (MUX) to produce WDM-encoded data streams. Each data stream gets routed by the AWGR to the different receivers depending on its wavelength, with any-to-any transmission achieved thanks to the frequency routing of the AWGR. At the receiving end (Rx) of each socket, the data stream gets demultiplexed with a 1:(N-1) optical demultiplexer (DEMUX) so that each wavelength is acquired by a separate socket, allowing to discriminate the transmission from each sender.

To counteract any wavelength or thermal instability and safeguard the functionality of the WDM-based system in real datacenters environments, feedback stabilization is required. The resonant wavelength of the microring based structures inside the photonic chip was stabilized thanks to the dithering technique, that implements

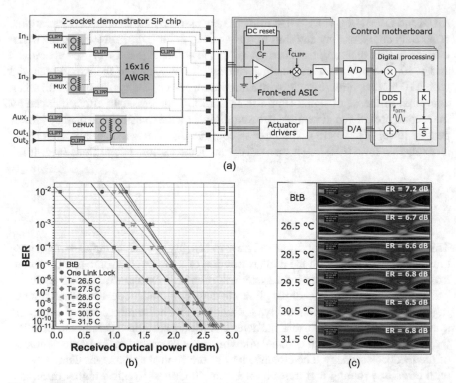

(a)

(b)

(c)

Fig. 4 **a** Silicon Photonic interconnection system integrated in a single chip and connected to the control electronics. **b** BER measurements and **c** eye diagrams at the output of the SiP chip, obtained at 30 Gbps at different temperatures

an easy yet effective power-and temperature-independent locking. To do so, a small sinusoidal signal is added to the MUX and DEMUX heater voltages, thus modulating the light in the output waveguide. The resulting light signal measured by the CLIPP is lock-in demodulated in the digital domain, obtaining, in this way, a voltage proportional to the first derivative of the transfer function of the device with respect to the temperature. Since the dithering signal is zero when the devices are at resonance, this information can be used as error signal to lock the rings.

The effectiveness of this approach has been experimentally demonstrated on a photonic chip featuring all the necessary building blocks to demonstrate optical communication between two transmitting sockets and a receiving one (Fig. 4a) [22]. A 30 Gbit s^{-1} modulated laser signal was injected into the system and monitored at the output to obtain both eye diagrams and BER measurements, while a second laser served as a crosstalk source. To simulate the operation of the chip in a thermally unstable environment, temperature oscillations were intentionally generated in the setup. In these conditions, the electronic feedback control applies to the heater of

MUX and DEMUX a voltage that mirrors very precisely the temperature variations, effectively compensating any shift of the working point. This is proved by the transmission quality measurements performed on the output signal. Error-free operations at 5×10^{-11} error-rate were obtained during all the 20 min of the experiment and within 10 °C temperature range, with a maximum power penalty of around 1 dB when compared to the input signal (Fig. 4b). In addition, the crosstalk effect induced by the second optical signal resulted to be negligible on the transmission quality (power penalty lower than 0.4 dB). The same is confirmed by the eye diagrams, that retain an extinction ratio (ER) larger than 6.5 dB in all the explored conditions (Fig. 4c).

6 Conclusions

This chapter presented and discussed how to reliably operate complex silicon photonic architectures through the implementation of a robust electronic feedback control. To guarantee scalability to the system, the non invasive nature of CLIPP detectors was exploited to monitor multiple photonic devices simultaneously without impairing the overall optical functionality. A complete low-noise mixed-signal electronic system was then designed to read the detectors with maximum accuracy and close the feedback loop in the digital domain, allowing great flexibility in the implementation of the feedback strategy. The effectiveness of the control system was demonstrated with error free routing and transmission of a 30 Gbit s^{-1} optical signal through a WDM-based routing engine. The result highlights that a proper electronic control layer can unlock the true potential of integrated optics and allow large-scale diffusion of this technology in real applications, paving the way to new sophisticated functionalities and architectures.

Acknowledgements The author acknowledges Marco Sampietro, Giorgio Ferrari and all the i3N lab for support during this work.

References

1. D.A.B. Miller, Rationale and challenges for optical interconnects to electronic chips. Proc. IEEE **88**(6), 728–749 (2000)
2. R.G. Beausoleil, Large-scale integrated photonics for high-performance interconnects. ACM J. Emerg. Technol. Comput. Syst. **7**(2), 1–54 (2011)
3. T. Alexoudi, N. Terzenidis, S. Pitris, M. Moralis-Pegios, P. Maniotis, C. Vagionas, C. Mitsolidou, G. Mourgias-Alexandris, G.T. Kanellos, A. Miliou, K. Vyrsokinos, N. Pleros, Optics in computing: from photonic network-on-chip to chip-to-chip interconnects and disintegrated architectures. J. Lightwave Technol. **37**(2), 363–379 (2019)
4. R. Bogue, Recent developments in MEMS sensors: a review of applications, markets and technologies, Sens Rev (2013)

5. S. Lloyd, M. Lim, The age of sensors—how MEMS sensors will enable the next wave of new products, in *2016 IEEE Symposium on VLSI Technology*, pp. 1–4 (2016)
6. A. Ribeiro, A. Ruocco, L. Vanacker, W. Bogaerts, "Demonstration of a 4 × 4-port self-configuring universal linear optical component, in *Progress in Electromagnetic Research Symposium (PIERS)* **2016**, 3372–3375 (2016)
7. K. Fotiadis, S. Pitris, M. Moralis-Pegios, C. Mitsolidou, P. De Heyn, J. Van Campenhout, T. Alexoudi, and N. Pleros, "16×16 silicon photonic AWGR for dense wavelength division multiplexing (DWDM) O-band interconnects, in *Proceeding SPIE OPTO 2020* (2020)
8. J. Sun, E. Timurdogan, A. Yaacobi, E.S. Hosseini, M.R. Watts, Large-scale nanophotonic phased array. Nature **493**(7431), 195–199 (2013)
9. S. Grillanda, V. Raghunathan, V. Singh, F. Morichetti, J. Michel, L. Kimerling, A. Melloni, A. Agarwal, Post-fabrication trimming of a thermal silicon waveguides. Opt. Lett. **38**(24), 5450–5453 (2013)
10. J. Michel, J. Liu, L.C. Kimerling, High-performance Ge-on-Si photodetectors. Nat Photon **4**(8), 527–534 (2010)
11. S. Assefa, F. Xia, Y.A. Vlasov, Reinventing germanium avalanche photodetector for nanophotonic on-chip optical interconnects. Nature **464**(7285), 80–84 (2010)
12. J.K. Doylend, A.P. Knights, The evolution of silicon photonics as an enabling technology for optical interconnection. Laser Photon. Rev. **6**(4), 504–525 (2012)
13. T.K. Liang, H.K. Tsang, I.E. Day, J. Drake, A.P. Knights, M. Asghari, Silicon waveguide two-photon absorption detector at 1.5 μm wavelength for autocorrelation measurements. Appl. Phys. Lett. **81**(7), 1323–1325 (2002)
14. T. Tanabe, H. Sumikura, H. Taniyama, A. Shinya, M. Notomi, All-silicon sub-Gb/s telecom detector with low dark current and high quantum efficiency on chip. Appl. Phys. Lett. **96**(10), 101103 (2010)
15. T. Baehr-Jones, M. Hochberg, A. Scherer, Photodetection in silicon beyond the band edge with surface states. Opt. Express **16**(3), 1659–1668 (2008)
16. H. Chen, X. Luo, A. W. Poon, Cavity-enhanced photocurrent generation by 1.55 μm wavelengths linear absorption in a pin diode embedded silicon microring resonator. Appl. Phys. Lett. **95**(17), 171111 (2009)
17. F. Morichetti, S. Grillanda, M. Carminati, G. Ferrari, M. Sampietro, M.J. Strain, M. Sorel, A. Melloni, Non-invasive on-chip light observation by contactless waveguide conductivity monitoring. IEEE J. Selected Top. Quantum Electron. **20**(4), 292–301 (2014)
18. F. Zanetto, E. Guglielmi, F. Toso, R. Gaudiano, F. Caruso, M. Sampietro, G. Ferrari, Wide dynamic range multichannel lock-in amplifier for contactless optical sensors with sub-ps resolution. IEEE Solid-State Circ. Lett. **3**, 246–249 (2020)
19. E. Guglielmi, M. Carminati, F. Zanetto, A. Annoni, F. Morichetti, A. Melloni, M. Sampietro, G. Ferrari, 16-channel modular platform for automatic control and reconfiguration of complex photonic circuits, in *IEEE International Symposium on Circuits and Systems (ISCAS)* **2017**, 1–4 (2017)
20. S. Grillanda, M. Carminati, F. Morichetti, P. Ciccarella, A. Annoni, G. Ferrari, M. Strain, M. Sorel, M. Sampietro, A. Melloni, Non-invasive monitoring and control in silicon photonics using cmos integrated electronics. Optica **1**(3), 129–136 (2014)
21. F. Zanetto, V. Grimaldi, F. Toso, E. Guglielmi, M. Milanizadeh, D. Aguiar, M. Moralis-Pegios, S. Pitris, T. Alexoudi, F. Morichetti et al., Dithering-based real-time control of cascaded silicon photonic devices by means of non-invasive detectors. IET Optoelectron. **15**(2), 111–120 (2021)
22. F. Zanetto, V. Grimaldi, M. Moralis-Pegios, S. Pitris, K. Fotiadis, T. Alexoudi, E. Guglielmi, D. Aguiar, P. De Heyn, Y. Ban et al., WDM-based silicon photonic multi-socket interconnect architecture with automated wavelength and thermal drift compensation. J. Lightwave Technol. **38**(21), 6000–6006 (2020)

Computer Science and Engineering

Beyond the Traditional Analyses and Resource Management in Real-Time Systems

Federico Reghenzani

Abstract The difficulties in estimating the Worst-Case Execution Time (WCET) of applications make the use of modern computing architectures limited in real-time systems. Critical embedded systems require the tasks of hard real-time applications to meet their deadlines, and formal proofs on the validity of this condition are usually required by certification authorities. In the last decade, researchers proposed the use of probabilistic measurement-based methods to estimate the WCET instead of traditional static methods. In this chapter, we summarize recent theoretical and quantitative results on the use of probabilistic approaches to estimate the WCET presented in the PhD thesis of the author, including possible exploitation scenarios, open challenges, and future directions.

1 Real-Time Systems and the WCET Problem

Real-time systems are computing systems in which the correctness of the computation does not depend only on the logic correctness—i.e., that the output is correctly produced—but also on the timing correctness—i.e., that the output is delivered within given time constraints. When such constraints must be satisfied at any time, and even a single violation is considered as a failure of the whole system, we call this system *hard* real-time. Vice versa, a *soft* real-time system allows the violation of timing constraints, provided that the violations do not occur *too often*.[1] Hard real-time systems are often *embedded systems*, and many of them are *mission-* or *safety-critical systems*. To mention a few examples: fly-by-wire computers of aircraft, the airbag control unit in a car, a pacemaker.

F. Reghenzani (✉)
Dipartimento di Elettronica, Informazione e Bioingegneria, Politecnico di Milano, Via Ponzio 34/5, 20133 Milano, Italy
e-mail: federico.reghenzani@polimi.it

[1] The term is voluntarily fuzzy. The exact definition of *too often* depends on the particular context.

© The Author(s) 2022
L. Piroddi (ed.), *Special Topics in Information Technology*,
PoliMI SpringerBriefs, https://doi.org/10.1007/978-3-030-85918-3_6

1.1 Scheduling Analysis

The system software plays a critical role in guaranteeing that the applications can satisfy timing requirements. In particular, the scheduler must be designed to correctly prioritize the tasks so that all of them meet their timing deadlines. This design is usually in contrast with general-purpose systems, where the focus is more on the total throughput of the system rather than the response-time of the single task.

During the design phase of a real-time system, the *scheduling analysis* verifies if a given scheduling algorithm is able to schedule all the tasks, correctly satisfying the timing constraints. Each task τ_i is represented, in the simplest task model, by a set of parameters $\tau_i = (T_i, C_i, D_i)$, where T_i is the period or inter-arrival time, C_i is the Worst-Case Execution Time (WCET), and D_i is the relative deadline. A task is an abstract entity periodically (with period T_i) or aperiodically (with minimum inter-arrival time T_i) activated. When a task is activated, it starts a new job. The job is the single unit of computation that performs the function the task is developed to. The job has a duration of maximum C_i and must complete its execution by the deadline D_i relative to the activation time.

Computing a correct Worst-Case Execution Time (WCET) is then essential to perform a legitimate scheduling analysis and the consequent claims of satisfying hard real-time constraints. The traditional way to estimate the WCET is to use the information on the hardware architecture combined with the software description (usually in the form of a control-flow graph) and derive the worst-case conditions leading to the WCET. It is not usually possible to compute the exact WCET, but only an approximation and, in particular, an over-estimation of it. A pessimistic over-estimation guarantees, in any case, a correct scheduling analysis.

1.2 The WCET Problem in Modern Architectures

Unfortunately, the evolution of hardware, particularly the processor, towards more complex computing architectures makes the computation of the WCET extremely difficult, or the estimated WCET is so pessimistic that it becomes unusable in practice. This is due to the advanced features added to respond to the increasing computational power demand of modern applications, such as machine learning, image vision, etc. To provide a trivial example, let us consider a multi-level cache hierarchy, very common in modern processors: forecasting a cache miss/hit of the single memory access of a program is non-trivial, and assuming all the memory accesses as miss makes the WCET extremely pessimistic (and the presence of cache substantially useless for the scheduling analysis standpoint).

The pervasive use of Commercial-off-the-Shelf (COTS) components in real-time applications is challenging because it adds another layer of complexity in WCET estimation, due to the numerous sources of unpredictability affecting these platforms [6]. In fact, COTS platforms are built with average performance in mind and are not intended to provide a timing model able to compute the WCET easily.

2 Probabilistic Real-Time Computing

A possible solution to the WCET estimation problem is the so-called *probabilistic real-time computing*: instead of a scalar value for the WCET, a statistical distribution is provided. This idea originates in the early 2000s from the papers by Edgar et al. [8] and Bernat et al. [1]. The statistical distribution can be estimated with two methods: a *Static Probabilistic Timing Analysis (SPTA)* and a *Measurement-Based Probabilistic Timing Analysis (MBPTA)*. The former estimates the distribution by looking at the same information available to the traditional static (but deterministic) analysis. However, it suffers the same problems of the deterministic analysis and, consequently, it did not spark too much interest in the scientific community. Vice versa, MBPTA is very attractive, thanks to its simplicity, and therefore it is the subject of this work. Two recent comprehensive surveys [5, 7] provide a general overview of probabilistic real-time WCET analyses research of the last years.

2.1 The Probabilistic-WCET

MBPTA approaches apply a statistical procedure to a finite sequence of random variables $X_1, X_2, \ldots, X_n$, representing the execution time of our task under analysis. The output of such procedures is a statistical distribution called *probabilistic-WCET (pWCET)*, and it is usually expressed with its Complementary Cumulative Distribution Function (CCDF):

$$p = P(X \geq \bar{C}) = 1 - F_X(\bar{C})$$

The probability p, called *violation probability*, represents the probability of observing an execution time larger than $\bar{C}$. The random variable X represents a generic random variable of the process by assuming that the random variables are identically distributed. The experimenter can select either $\bar{C}$ or p and accordingly compute the other value: it is possible to estimate the violation probability p given a WCET $\bar{C}$ or, vice versa, estimate the WCET $\bar{C}$ given a target violation probability p. The latter option is computed by using the Inverse Cumulative Distribution Function (ICDF).

Provided that the pWCET distribution is correctly computed and it represents the real distribution of execution times, it is reasonable to claim that we are compliant with safety-critical processes: a probability of violation, corresponding to the real one, would be just another term in the failure analysis of safety-critical systems. However, guaranteeing that the probability of violation is correctly computed is non-trivial and represents the major obstacle to probabilistic real-time use in the current industrial system. We will discuss this issue in Sect. 3.

2.2 Extreme Value Theory

The statistical theory called *Extreme Value Theory (EVT)* provides a reliable way to estimate the pWCET from the observations of the execution time. The mathematical details on this statistical theory are omitted here due to space limitations, but it can be found in the thesis [10] or in specialized books [4]. The main EVT result is the *Fisher-Tippett-Gnedenko theorem* which states that the distribution tail of an observed phenomenon—i.e., in our case, the maxima of execution times—converges to the Gumbel, the Weibull, or the Fréchet distribution, independently from the original distribution of the measured values. This is a key property because we can estimate the pWCET without knowledge of the original distribution of the execution times. Moreover, the three distributions are actually particular cases of a more general distribution: the *Generalized Extreme Value Distribution (GEVD)*, which is characterized by the following Cumulative Distribution Function (CDF):

$$G(x) = \begin{cases} e^{-e^{\frac{x-\mu}{\sigma}}} & \xi = 0 \\ e^{-[1+\xi(\frac{x-\mu}{\sigma})]^{-1/\xi}} & \xi \neq 0 \end{cases} \tag{1}$$

The GEVD distribution $\mathscr{G}(\mu, \sigma, \xi)$ is parameterized by the *location* parameter μ, the *scale* parameter σ, and the *shape* parameter ξ. The latter determines which subclass the distribution is ($\xi = 0$: Gumbel, $\xi < 0$: Weibull, or $\xi > 0$: Fréchet). Equivalently, it is possible to use the *Generalized Pareto Distribution (GPD)*. This theory is applicable provided that three conditions are satisfied:

- The input measurements are *independent and identical distributed (i.i.d.)*;
- The real distribution is in the *Maximum Domain of Attraction (MDA)* of an EVT distribution;
- The measurements used in analysis are *representative* of the real execution.

The next section focuses on explaining how to verify that these conditions are valid.

3 Uncertainty Estimation

The estimation of the pWCET is performed via a proper set of algorithmic steps. The overall process leading to estimate the pWCET distribution is, in fact, more sophisticated than just running a distribution estimator. In particular:

1. The sequence of execution time measurements $X_1, X_2, \ldots, X_n$ is tested to verify the validity of the i.i.d. hypothesis;
2. A filtering technique is applied to the time measurements to capture only the "tail part" of th distribution;
3. The remaining samples are used to feed a distribution estimator (such as Maximum Likelihood Estimator or Probabilistic Weighted Moment);

4. A Goodness-of-Fit (GoF) test is performed to verify the correspondence between the estimate distribution and the original set of samples (implicitly verifying also the MDA hypothesis).

3.1 The Importance of Statistical Testing

In the context of the aforementioned estimation process, we can distinguish two types of statistical tests we need: (1) a test to verify the i.i.d. hypothesis and (2) the GoF test for the final check of the distribution. However, statistical testing is subject to errors due to the obvious finiteness of the input measurements, and this may impact the final reliability of the obtained pWCET [15]. For this reason, we created two mathematical tools that help an experimenter to assess the quality and reliability of the obtained pWCET: The Probabilistic Predictability Index to check the i.i.d. hypothesis and the Region of Acceptance to verify the Goodness-of-Fit test.

3.2 The Probabilistic Predictability Index

The requirement of the i.i.d. hypothesis is actually stricter than needed, and it is possible to split it into three sub-hypothesis [17]: stationarity, short-range independence, and long-range independence. For each of these three categories, we selected statistical tests capable of identifying a violation in these properties [12, 14]: KPSS, BDS, and R/S tests. These tests can be used to evaluate the ability of hardware and software to comply with the three sub-hypotheses of the i.i.d. hypothesis. However, comparing different solutions using separate tests is non-trivial, mainly due to the effect on the significance level α. For this reason, we developed an index called *Probabilistic Predictability Index (PPI)* [12], which maintains the statistical properties of the original tests while providing a convenient way to compare hardware/software solutions.

The PPI is a number in the range (0, 1) calculated with the following equation:

$$PPI := \begin{cases} \min_{\forall i} f_i(D_i) \cdot \prod_{i \in v^*}[1 - (CV_{PPI} - f_i(D_i))] & v \neq \emptyset \\ \frac{1}{3}\sum_{\forall i} f_i(D_i) & v = \emptyset \end{cases} \quad (2)$$

where

- f_i are the following functions:

  - $f_{KPSS}(x) = e^{-K_{KPSS} \cdot x}$;
  - $f_{BDS}(x) = e^{-K_{BDS} \cdot |x|}$;
  - $f_{R/S}(x) = e^{-K_{R/S} \cdot x}$
  - with $k_{KPSS} = \frac{1}{4}$, $k_{BDS} = -\frac{\log f_{KPSS}(CV_{KPSS})}{|CV_{BDS}|}$, $k_{R/S} = -\frac{\log f_{KPSS}(CV_{KPSS})}{CV_{R/S}}$

- D_i is the value of the *statistic* of each test computed via the original formulas for the KPSS, BDS, and R/S tests;

Fig. 1 An example of region
of acceptance depicted with
the three axes matching the
three GEV parameters

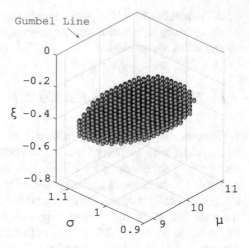

- CV_{PPI} is the critical value for the PPI test, computed as $CV_{PPI} = f_{KPSS}(CV_{KPSS})$
- v is the violation set, i.e. $v = \{i \mid f_i(D_i) < CV_{PPI}\}$, and v^* is the violation set without the minimum, i.e. $v^* = \{v \setminus \arg\min_{\forall i} f_i(D_i)\}$.

A full description of the steps to derive this formulation is available in the thesis [10]. When this number approaches 1, then the input time series appears to be compliant with the i.i.d. hypothesis, while when approaches 0 the series is non-compliant. The decision value is set at CV_{PPI}.

3.3 Region of Acceptance

The output of the estimator routine of step 3 of the EVT process is the set of parameters of our distribution. Let us write $(\overline{\mu}, \overline{\sigma}, \overline{\xi})$ the tuple of estimated parameters. The tuple $(\mu^{\circledast}, \sigma^{\circledast}, \xi^{\circledast})$ is the exact, but unknown, distribution. The goal of the GoF test of step 4 of the EVT process is to identify whether the distance between these two tuples is *too large* to compromise the safety of the final pWCET. The output of the GoF test is a region in the parameters space identifying the limits of the acceptable distribution parameters. We call this cloud of point, depicted in Fig. 1, *Region of Acceptance (RoA)*. If the estimated parameter tuple $(\overline{\mu}, \overline{\sigma}, \overline{\xi})$ is inside this region, then the pWCET distribution can be safely accepted, according to the confidence provided by the test.

We know [16, Theorem 3.6] that the exact distribution $(\mu^{\circledast}, \sigma^{\circledast}, \xi^{\circledast})$ is inside the RoA or at its border. For this reason, even considering the limitation of the estimator and the statistical test, we can find a distribution that over-estimates all the others by looking at the boundaries of this multi-dimensional space. Therefore, given a region R, a violation probability $\overline{p}$, and a point $\hat{P} \in R$ such that $\hat{P}$ is the point that maximizes the CCDF of the region, then either $\hat{P} = (\mu^{\circledast}, \sigma^{\circledast}, \xi^{\circledast})$ or the pWCET associated to $\hat{P}$ overestimates the real pWCET given by $(\mu^{\circledast}, \sigma^{\circledast}, \xi^{\circledast})$ at violation probability $\bar{p}$.

The described result is only a first step in the analysis of the RoA. Then, it is possible to derive several mathematical procedures to explore the RoA (described in [16]), dealing with the trade-off pessimism and reliability of the final pWCET result.

4 Exploiting Probabilistic Real-Time

Probabilistic real-time is still affected by numerous open challenges and problems (subsequently explained in Sect. 5) to consider it ready to use for WCET estimation of tasks in safety-critical systems. However, we can already trust it in other cases, such as Mixed-Criticality, High-Performance Computing, or to estimate the worst-case energy. The following paragraphs briefly describe these three scenarios and how probabilistic real-time plays a role in them.

4.1 Mixed-Criticality

Mixed-Criticality Systems are systems providing a mix of critical functions not all at the same criticality level. In the context of real-time computing, the traditional task model is the one proposed by Vestal in 2007 [18]. Multiple values of WCET are assigned to each task depending on its criticality: higher criticality tasks have several WCETs, depending on the level of assurance used to estimate it. The WCET with the highest level of assurance is usually the one computed with the traditional static method, which is safe but very pessimistic. We can exploit probabilistic information in mixed-criticality systems to: (1) estimate the non-highest level of assurance WCETs with reasonable accuracy, and (2) improve non-functional requirements (such as energy consumption) while still guaranteeing the hard deadline with the statically computed WCET.

In the first case, the idea is to exploit the pWCET to estimate the values of C_i^j, which is the execution time of the task τ_i at the assurance level j. At the highest assurance level ($j = $ HI), C_i^{HI} is computed with safe static methods. Then, for lower assurance levels, we can use the pWCET by setting different values for the violation probability p. Even in the case of the pWCET being incorrect, at least the correctness of HI-criticality level tasks is guaranteed. For further details on mixed-criticality scheduling, refer to the Burns et al. survey [3].

In the second case, the timing properties are guaranteed with the WCET estimated with static analysis, while secondary properties, such as energy consumption, temperature, etc., are optimized based on probabilistic information. In this way, an incorrect estimation of the distribution does not invalidate the safety properties. This is an active area of research and subject of a recent work [2].

4.2   High-Performance Computing

High-Performance Computing (HPC) clusters are composed of hundreds or thousands of general-purpose servers. What is the relation with embedded safety-critical systems we are talking about in this chapter? Because time-critical applications are also emerging on these systems [13], such as medical imaging, natural disaster prediction, or structures monitoring. All of these application categories require a large amount of computational power (consequently the need to run on HPC clusters) and timing guarantees on the results. Clearly, the deadlines, in this case, are orders of magnitude larger than the deadline in the embedded system case, but the problem of scheduling real-time workload remains the same.

The static computation of WCET on HPC hardware and software is practically impossible. This is due to the complex general-purpose architecture of the single machine and of the network. Exploiting MBPTA and the resulting pWCET can solve the problem because it does not require perfect modeling of hardware and software. However, the EVT hypotheses must be satisfied. First preliminary results [13] showed that the EVT hypotheses could be satisfied with due safety technical shrewdness. The presence of heterogeneous hardware (such as GPGPU computing), which is exploding in HPC in the last years, exhibited an improvement in compliance of EVT hypotheses compared to a full-homogeneous scenario.

4.3   Energy Estimations

Even more difficult than the WCET problem, it is the *Worst-Case Energy Consumption (WCEC)* problem. The WCEC is necessary for some critical systems having an energy budget to satisfy as a *functional* requirement. Typical scenarios include systems powered by energy-harvesting devices (e.g., solar panels), such as embedded systems located in remote regions not having access to the power grid or satellites harvesting energy only when exposed to the sun. In these situations, the WCEC is needed to formally verify that the system can survive the period without a stable energy source. However, estimating the WCEC requires not only the WCET estimation as input but also the perfect model of the hardware in terms of power consumption.

To overcome the WCEC estimation problem, we proposed [11] to use the same theory used for the pWCET but directly applied to energy (or power) measurements of our system. In this way, similarly to MBPTA, we can hide the complexity of the power/energy model of the system and exploits EVT to obtain a pWCEC estimation. The same pWCET limitations and hypotheses, which must be satisfied, exist. However, differently from the WCET case, in the WCEC case, the choice of pWCEC is almost mandatory due to the difficulties in estimating even a pessimistic static WCEC.

5 Current Open Challenges and Future Directions

In the previous section, we discussed the i.i.d. and MDA hypotheses and how to verify them with statistical testing. We have not yet discussed about the third EVT hypothesis: representativity. The *representativity* informally means that we observed a *sufficient amount* of application behaviors to be sure we minimize the epistemic uncertainty of the phenomenon under statistical analysis. For example, suppose the control-flow graph of a program under analysis has a branch that is never taken during the measurement campaign. In that case, no statistical technique can infer something it cannot see. Representativity is the major obstacle in certifying safety-critical software by exploiting probabilistic real-time. The presence of the probability itself is not a risk for the safety, provided that the probability is perfectly computed. In such a scenario, the probability is added as another term of the failure analysis (like a hardware failure). A more in-depth discussion on representativity is available in the thesis [10].

Besides representativity, several other challenges related to probabilistic real-time are open [9], especially on uncertainty estimation and how to build hardware architectures able to comply with the EVT hypotheses. Therefore, the research is still very active and presents numerous challenges to address in the next years.

In particular, we identified three research fronts we believe to be promising subjects of future research on probabilistic real-time for the next years:

- Continuing the study of the theory behind the pWCET and its safety, with a particular focus on the representativity problem and uncertainty estimation;
- How to exploit, in a different manner, the current probabilistic real-time theory. This includes a more in-depth analysis of pWCET in HPC clusters, the optimization of non-functional metrics (such as energy, power, reliability, temperature, etc.), and monitoring application behaviors via statistical techniques;
- Dealing with fault-tolerance requirements, for example, by allowing tasks to re-execute if a failure occurs. In this scenario, the pWCET information can be exploited to verify the probability of transient faults to happen and perform probabilistic scheduling on the task re-execution.

The scientific community is divided over the future of probabilistic real-time: the barrier of representativity is seen as insuperable for many. However, static WCET analyses are also stuck for years. Our opinion is that it is worth continuing to investigate the probabilistic theory, in particular to quantify how much we can rely on its output—i.e., the pWCET—and to discover other use-cases of the probabilistic information.

References

1. G. Bernat, A. Colin, S.M. Petters, WCET analysis of probabilistic hard real-time systems, in *23rd IEEE Real-Time Systems Symposium, 2002. RTSS 2002* (IEEE, 2002), pp. 279–288
2. A. Bhuiyan, F. Reghenzani, W. Fornaciari, Z. Guo, Optimizing energy in non-preemptive mixed-criticality scheduling by exploiting probabilistic information. IEEE Trans. Comput.-Aided Des. Integrated Circ. Syst. **39**(11), 3906–3917 (2020)
3. A. Burns, R. Davis. Mixed criticality systems-a review. *Department of Computer Science, University of York, Tech. Rep* (2013), pp. 1–69
4. E. Castillo, Extreme Value Theory in Engineering (Elsevier, Statistical Modeling and Decision Science, 2012)
5. F.J. Cazorla, L. Kosmidis, E. Mezzetti, C. Hernandez, J. Abella, T. Vardanega, Probabilistic worst-case timing analysis: taxonomy and comprehensive survey. *ACM Comput. Surv.*, 52(1):14:1–14:35, February 2019
6. D. Dasari, B. Akesson, V. Nélis, M.A. Awan, S.M. Petters. Identifying the sources of unpredictability in COTS-based multicore, in *International Symposium on Industrial Embedded Systems* (IEEE, 2013), pp. 39–48
7. R. Davis, L. Cucu-Grosjean, A survey of probabilistic timing analysis techniques for real-time systems. Leibniz Trans. Embedded Syst. **6**(1), 03–1–03:60 (2019)
8. S. Edgar, A. Burns, Statistical analysis of WCET for scheduling, in *Proceedings 22nd IEEE Real-Time Systems Symposium (RTSS 2001) (Cat. No.01PR1420)* (2001), pp. 215–224
9. S. Jiménez Gil, I. Bate, G. Lima, L. Santinelli, A. Gogonel, L. Cucu-Grosjean, Open challenges for probabilistic measurement-based worst-case execution time. IEEE Embedded Syst. Lett. **9**(3), 69–72 (2017)
10. F. Reghenzani, *Beyond the Traditional Analyses and Resource Management in Real-Time Systems*. PhD thesis, Politecnico di Milano, Jan 2021. Advisor: Prof. William Fornaciari
11. F. Reghenzani, G. Massari, W. Fornaciari, A probabilistic approach to energy-constrained mixed-criticality systems, in *2019 IEEE/ACM International Symposium on Low Power Electronics and Design (ISLPED)* (2019), pp. 1–6
12. F. Reghenzani, G. Massari, W. Fornaciari, Probabilistic-WCET reliability: statistical testing of EVT hypotheses. Microproc. Microsyst. 77, 103135 (2020)
13. F. Reghenzani, G. Massari, W. Fornaciari, Timing predictability in high-performance computing with probabilistic real-time. IEEE Access 8, 208566–208582 (2020)
14. F. Reghenzani, G. Massari, W. Fornaciari, A. Galimberti, Probabilistic-WCET reliability: on the experimental validation of EVT hypotheses, in *Proceedings of the International Conference on Omni-Layer Intelligent Systems*, COINS '19, New York, NY, USA (2019), pp. 229–234. Association for Computing Machinery
15. F. Reghenzani, L. Santinelli, W. Fornaciari, Why statistical power matters for probabilistic real-time: work-in-progress, in *Proceedings of the International Conference on Embedded Software Companion*, EMSOFT '19, New York, NY, USA (2019). Association for Computing Machinery
16. F. Reghenzani, L. Santinelli, W. Fornaciari, Dealing with uncertainty in pWCET estimations. ACM Trans. Embed. Comput. Syst. **19**(5) (2020)
17. L. Santinelli, J. Morio, G. Dufour, D. Jacquemart, On the sustainability of the extreme value theory for WCET estimation, in *14th International Workshop on Worst-Case Execution Time Analysis*, volume 39 of *OpenAccess Series in Informatics (OASIcs)*, pages 21–30, Germany, 2014. Schloss Dagstuhl–Leibniz
18. S. Vestal, Preemptive scheduling of multi-criticality systems with varying degrees of execution time assurance, in *28th IEEE International Real-Time Systems Symposium (RTSS 2007)* (2007), pp. 239–243

Computational Inference of DNA Folding Principles: From Data Management to Machine Learning

Luca Nanni

Abstract DNA is the molecular basis of life and would total about three meters if linearly untangled. To fit in the cell nucleus at the micrometer scale, DNA has, therefore, to fold itself into several layers of hierarchical structures, which are thought to be associated with functional compartmentalization of genomic features like genes and their regulatory elements. For this reason, understanding the mechanisms of genome folding is a major biological research problem. Studying chromatin conformation requires high computational resources and complex data analyses pipelines. In this chapter, we first present the PyGMQL software for interactive and scalable data exploration for genomic data. PyGMQL allows the user to inspect genomic datasets and design complex analysis pipelines. The software presents itself as a easy-to-use Python library and interacts seamlessly with other data analysis packages. We then use the software for the study of chromatin conformation data. We focus on the epigenetic determinants of Topologically Associating Domains (TADs), which are region of high self chromatin interaction. The results of this study highlight the existence of a "grammar of genome folding" which dictates the formation of TADs and boundaries, which is based on the CTCF insulator protein. Finally we focus on the relationship between chromatin conformation and gene expression, designing a graph representation learning model for the prediction of gene co-expression from gene topological features obtained from chromatin conformation data. We demonstrate a correlation between chromatin topology and co-expression, shedding a new light on this debated topic and providing a novel computational framework for the study of co-expression networks.

1 Introduction

DNA is the molecular basis of life. It stores all the information required for life to reproduce, sustain itself and adapt to the environment. The human genome con-

L. Nanni (✉)
Dipartimento di Elettronica, Informazione e Bioingegneria, Politecnico di Milano, Via Ponzio 34/5, 20133 Milano, Italy
e-mail: luca.nanni@polimi.it; luca.nanni@unil.ch

Department of Computational Biology, Université de Lausanne, Génopode, 1015 Lausanne, Switzerland

© The Author(s) 2022
L. Piroddi (ed.), *Special Topics in Information Technology*,
PoliMI SpringerBriefs, https://doi.org/10.1007/978-3-030-85918-3_7

sists of approximately three billion DNA base pairs [1, 2], which would total about three meters if linearly untangled. Therefore, to fit in the cell nucleus, which has an approximate diameter of six micrometers, DNA has to fold itself several times.

The study of DNA folding is a current biological research frontier and only recently it was possible to reliably probe its characteristics. A major technological milestone, which enabled for the first time to see the whole genome three-dimensional organization, was the invention of the Hi-C technique [3], which maps 3D contacts between all genomic locations. For the first time it was possible to reveal that genomes fold into high-level structures called Topologically Associating Domains (TADs) [4], which are genomic regions of high self-interaction and low interactions across them. TADs are of great biological relevance, since they are speculated to be associated with functional compartmentalization of genomic features like genes and their regulatory elements. For this reason, understanding the mechanisms of TADs and chromatin loops formation is a major biological research problem.

Studying chromatin conformation and Hi-C data requires high computational resources and complex data analyses pipelines. This is due to both the intrinsic noise present in the data and to the big size of the experimental datasets [5]. Hi-C experiments can produce billions of reads and their storage can require several terabytes. For this reason, it is necessary to develop scalable and interactive software for the exploration of big genomic datasets like Hi-C. In addition, studying Hi-C data usually requires the integration of other heterogeneous data sources (like, for example, ChIP-seq and RNA-seq), which implies that any software pipeline designed for complex biological analyses on chromatin conformation has to be able to work with multiple data formats as well as be able to easily integrate and manipulate them.

In the first part of this chapter, we present the design and implementation of a software for interactive and scalable data exploration for genomic data. This system, named PyGMQL, is a easy-to-use Python extension of the GMQL big genomic data engine, developed by the Genomic Computing team at Politecnico di Milano. PyGMQL is a Python package which enables the user to build complex genomic pipelines by relying on the Spark big data engine. Using a carefully designed set of functions, it allows the user to inspect and manipulate arbitrarily big genomic datasets and design the flow of execution of the program. Once the researcher has defined the complete data analysis pipeline, it can then be fully executed on the complete dataset. It also allows the loading and manipulation of heterogeneous data, coming from different experimental procedures, through the adoption of the Genomic Data Model (GDM), developed initially for the GMQL system. We demonstrate the qualities and performance of this software through a series of data analysis scenarios, showing its usability, requirements and execution times. PyGMQL is extremely versatile and can be used both a downstream analysis data analysis tool and as a development framework on top of which bioinformaticians can build new specialized software tools.

In the second part, we apply this software stack to the study of chromatin conformation data. We focus on the study of the epigenetic drivers and determinants of TADs, and we explore the relationship between the positions and motif orientations of CTCF binding sites. CTCF is a structural protein known to be associated with chromatin looping [6], but the underlying rules determining TAD topologies have never been highlighted. We propose a set of spatial rules, having as elements the binding sites of CTCF, which correlate with TADs and their topological characteristics. We argue that genome conformation can be explained by a set of "grammatical rules" acted out by CTCF.

We finally inspect the relationship between chromatin conformation and gene expression. We ask to what extent chromatin networks determined by Hi-C can explain co-expression relations between genes, both intra- and inter-chromosomically. We model this research question as a machine learning problem, designing a graph representation learning model for the encoding of gene chromatin topological features. The learnt gene embeddings are used then as inputs to a Random Forest classifier. We demonstrate that this model outperforms a set of baselines based on previous works.

2 Interactive and Scalable Data Analysis for Genomics

The analysis of biological data is conventionally divided into three macro-steps, depending on the data processing and the biological question being asked and answered. This categorization is particularly important in the case of next-generation sequencing data.

- **Primary data analysis**: The first step in the analysis of biological data is usually the production and quality assessment of the raw data generated from the sequencing machine. It produces the raw nucleotide sequences for each read obtained from the machine.
- **Secondary data analysis**: Reads are then filtered and *aligned* to a *reference genome*, which gives the researcher the information about the *position* of a sequencing read in the target genome.
- **Tertiary data analysis**: Researchers then deepen into the biological problem they are studying and ask high-level questions about the data. This is the most important and case-specific step of biological data analysis and it usually requires custom pipelines and result evaluation, as well as the integration of different data sources.

2.1 The Genometric Query Language and Its Ecosystem

The Genomic Computing (GeCo) team at Politecnico di Milano decided to address these issue of *managing, extracting and analysing* tertiary biological data by proposing a paradigm shift in genomic data management [7].

The *Genomic Data Model* (GDM) represents genomic information as the combination of *region data* and *metadata* [8]. The *Genometric Query Language* (GMQL) [7, 9] was designed to answer complex biological questions using a declarative query language. GMQL queries are compiled into a *computation graph*. This enables the implementation of different *backends* of the GMQL system [9]. GMQL has support for Spark [10], Flink [11] and SciDB [12] implementations. The GMQL repository [13] hosts data coming from The Cancer Genome Atlas [14], Roadmap Epigenomics [15], ENCODE [16] and others. The main interface of the GMQL system is through a *web application*, where the user can browse through the publicly available datasets in the GMQL repository, write and compile genomic queries through the *query editor*, and inspect the results of queries in his/her private storage.

2.2 PyGMQL: Scalable Programmatic Data Analysis of Genomic Data

[1] GMQL operates in a batch fashion, which means that the scientists has to write the query, run it and wait for the outcome. This approach does not fit very well the data exploration paradigm. This is due the fact that, in particular at the beginning of a study, scientists do not know the specific research question they want to ask to their data. This approach requires a shift of our framework towards *interactive* computation, where parameters are learnt during the exploration process through trial and error.

With the aim of solving this issue, we designed and implemented PyGMQL [17, 18], a Python library which embeds the GMQL engine.

The library adopts a client-server architecture, where the Python front-end exposes to the user all the dataset manipulation functions and utilities, while a Scala back-end implements all the query operators. As depicted in Fig. 1a, the back-end relies on the implementation of GMQL on Spark.

PyGMQL offers a set of methods which wrap and extend the GMQL language operators. PyGMQL offers methods to manipulate genomic datasets as variables, each associated to structures called GMQLDataset. The library stores all the operations performed on a variable through a GMQL directed acyclic graph (DAG) (see [9]). This design enables the back-end to apply query optimizations to the DAG structure [19]. Conforming to the *lazy execution* model of Spark, no actual operation is executed until the user explicitly triggers the computation.

PyGMQL can access data stored in the local machine, like most of the Python libraries. In addition, PyGMQL can also interface with an external GMQL system to interact with his/her private datasets or the public repository. Therefore, queries in PyGMQL can be composed of genomic operations acting both on local and remote datasets. The library can also "outsource" the query computation to an external GMQL service, and then download the result to the user machine (Fig. 1b). The location of the datasets used during the query is therefore orthogonal with respect to the mode of execution. The library keeps tracks of the used datasets and their dependen-

---

[1] This section is adapted from [17].

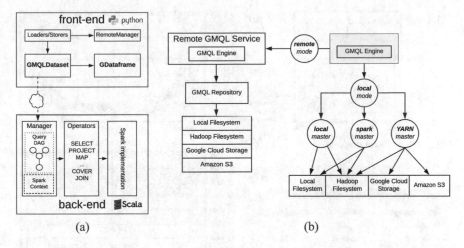

(a)   (b)

Fig. 1 **a** Schematic representation of the software components of PyGMQL. In the front-end, the `GMQLDataset` is a data structure associated with a query. The `GDataframe` stores the query result and enables in-memory manipulation of the data. The `RemoteManager` module is used for message interchange between the package and an external GMQL service. The back-end is implemented in Spark. **b** Deployment modes and executor options of the library. Figure adapted from [17]

cies during the whole Python program execution, minimizing the data transmission between local and remote systems.

We demonstrated the flexibility of the PyGMQL library through three data analysis workflows, available in the PyGMQL GitHub repository.[2] For a progressive introduction to PyGMQL usage, the applications are increasingly complex both for what concerns the biological analysis and the data size. We also provide a performance evaluation on increasingly large datasets of a significantly complex genomic query, highlighting the scalability of the system [17].

3 The Grammar of Genome Folding

[3] We then decided to apply our novel software stack to a concrete biological research problem, having as focus the study of the mechanisms of chromatin folding.

The mechanisms underlying the storage and packing of DNA is a current knowledge frontier in biology. Thanks to the Hi-C technique it was possible to identify genomic regions of having an high level of self-interaction, which were named Topologically Associating Domains (TADs) [4]. TADs are divided by boundaries that are detected as positions in the genome where there is a sharp break from preferential

---

[2] https://github.com/DEIB-GECO/PyGMQL.

[3] This section is adapted from [20].

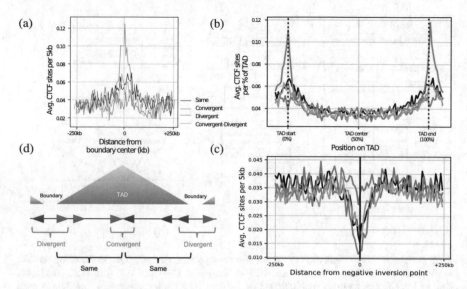

Fig. 2 **a** Enrichment of the four CTCF pattern classes around highly conserved TAD boundaries in 7 cell types. **b** Enrichment of the same classes along the length of TADs, as a function of percentage of TAD size. **c** Enrichment of the same classes around the center of TADs. **d** Schematic representation of the "grammatical rules" induced by CTCF and its orientation, and their correlation with TAD features. Figure adapted from [20]

left-ward interactions to preferential right-ward interactions [4, 21]. Scientists also tried to understand the underlying factors determining chromatin topological organization. It was shown that the CTCF protein is a fundamental actor in this context. The CTCF protein binds to an asymmetric motif on the DNA [22]. This means that it is possible to find these motifs in two possible *orientations* (> and <). It has been shown that convergent CTCF binding sites (> <) are located at the two extremities of long range interacting DNA regions (loops), explaining multiple features of Hi-C data sets [6, 23].

With the aim to reconcile the notion of chromatin loops and TADs, we created a classification scheme for sets of more than two adjacent CTCF sites. Each cluster is classified based on the relative orientation of the CTCF binding sites composing it[20]. We identified four categories: *same* (>>>, <<<), *convergent* (>><, > <<), *divergent* (<>>, <<>) and *convergent-divergent* (> <>, <> <). We therefore analysed the distributions of the sizes of CTCF clusters the human genome, revealing that, at length scales ranging from 5 to 100 kb, divergent CTCF site clusters are enriched, while convergent CTCF site clusters are depleted. This suggests that divergent CTCF sites code for TAD boundaries (Fig. 2a, b) and that convergent CTCF sites are involved in the definition of left and right TAD sections (Fig. 2c). We validated this orientation-based grammar as a function of CTCF site strength and TAD boundary strength (Fig. 2a).

These results suggest the presence of a linear "grammar" dictating the relative orientations of CTCF binding sites motifs, and that these simple rules play an impor-

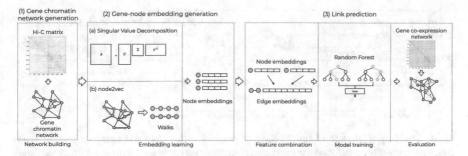

Fig. 3 Schematic representation of the workflow. We generate a Gene Chromatin Network by summarizing Hi-C information of genes and their neighborhood for each gene (**1**). We then produce a reduced vector representations of genes through network embedding techniques (**2**). Finally, we take the combined pairs of gene vectors as input for a Random Forest classifier, trained on a subset of the gene co-expression network (**3**). Figure adapted from [30]

tant role in the definition of key chromatin conformation features, like TADs, their inside and their boundaries (Fig. 2d).

4   Chromatin Conformation and Gene Expression

A key element left out from our previous analysis of chromatin conformation is the role of genes. Gene expression is the key biological mechanism which produces phenotypic differences across cell types, tissues and biological conditions. Gene expression data can be summarized through the computation of co-expression between pairs of genes, building therefore a *co-expression network*.

The relationship between *gene co-expression* and chromatin conformation is of great biological interest [24–26]. Given the high complexity of Hi-C data and the difficult definition of gene coexpression networks [27], the development of proper computational tools to investigate such relationship is rapidly gaining the interest of researchers. One of the most fascinating questions in this context is how chromatin topology correlates with gene coexpression and which physical interaction patterns are most predictive of coexpression relationships [28, 29].

We explored the relationship between chromatin conformation and gene expression using a predictive modeling approach. Specifically, we designed a model to predict co-expression between two genes from the physical set of interactions derived from an Hi-C experiment (see Fig. 3). We used a representation learning approach [31] for embedding the topological features of genes. In our work, the features of the nodes are learnt by solving an optimization problem, which defines the embedding strategy of the physical interaction network extracted from Hi-C data. Therefore, the proper choice of the optimization method is critical.

We compared two different node embedding strategies. The first method is based on Matrix Factorization [32], while the second exploits a random walk procedure to

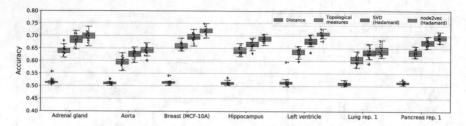

Fig. 4 Accuracy measure across the 22 single-chromosome networks of the proposed gene embedding strategies and comparison with baselines. Each box plot is derived from the accuracy measures for each cross validation fold and for each chromosome. Figure adapted from [30]

find similar embeddings for genes in the same neighborhood [33]. We then used the learnt embeddings to train a non-linear classifier, based on Random Forest [34]. We compared the performances between the two embedding strategies and against a set of baselines: the first one is based on the computation of a set of predefined measures for each gene/node in the Hi-C network to be used as input of the classifier [24]; the second is a simple distance-based predictor using only the linear distance between genes in the genome as input.

We validated our models on a comprehensive collection of datasets, where Hi-C data and matching gene expression is available, showing consistent results across conditions, cell line and tissues. Our results (see Fig. 4) show that both our embedding methods outperform the considered baselines. This finding shows that it is indeed possible to encode the topological properties of genes, and to use them to predict their co-expression. It is also important to point out that, despite the significant improvement in prediction performance of our proposed methods, we were able to predict correctly only a portion of the co-expression relationships. A possible interpretation of this result is that chromatin conformation and gene expression are linked phenomena, but they also maintain a degree of independence: the gene expression machinery is a very complex system, influenced by several factors, like histone modifications and DNA methylation, together with chromatin conformation.

References

1. International Human Genome Sequencing Consortium et al. Initial sequencing and analysis of the human genome. Nature, **409**(6822), 860 (2001)
2. J.C. Venter, M.D. Adams, E.W. Myers, P.W. Li, R.J. Mural, G.G. Sutton, H.O. Smith, M. Yandell, C.A. Evans, R.A. Holt et al., The sequence of the human genome. Science **291**(5507), 1304–1351 (2001)
3. E. Lieberman-Aiden, N.L. Van Berkum, L. Williams, M. Imakaev, T. Ragoczy, A. Telling, I. Amit, B.R. Lajoie, P.J. Sabo, M.O. Dorschner et al., Comprehensive mapping of long-range interactions reveals folding principles of the human genome. Science **326**(5950), 289–293 (2009)
4. J.R. Dixon, S. Selvaraj, F. Yue, A. Kim, Y. Li, Y. Shen, M. Hu, J.S. Liu, B. Ren, Topological domains in mammalian genomes identified by analysis of chromatin interactions. Nature **485**(7398), 376 (2012)

5. B.R. Lajoie, J. Dekker, N. Kaplan, The hitchhiker's guide to hi-c analysis: practical guidelines. Methods **72**, 65–75 (2015)
6. A.L. Sanborn, S.S.P. Rao, S.-C. Huang, N.C. Durand, M.H. Huntley, A.I. Jewett, I.D. Bochkov, D. Chinnappan, A. Cutkosky, J. Li et al., Chromatin extrusion explains key features of loop and domain formation in wild-type and engineered genomes. Proc. Nat. Acad. Sci. **112**(47), E6456–E6465 (2015)
7. M. Masseroli et al., Genometric query language: a novel approach to large-scale genomic data management. Bioinformatics **31**(12), 1881–1888 (2015)
8. M. Masseroli et al., Modeling and interoperability of heterogeneous genomic big data for integrative processing and querying. Methods **111**, 3–11 (2016)
9. M. Masseroli et al. Processing of big heterogeneous genomic datasets for tertiary analysis of next generation sequencing data. Bioinformatics, pp bty688 (2018)
10. M. Zaharia et al., Apache spark: a unified engine for big data processing. Commun. ACM **59**(11), 56–65 (2016)
11. A. Katsifodimos, S. Schelter. Apache flink: stream analytics at scale, in *2016 IEEE International Conference on Cloud Engineering Workshop (IC2EW)*. IEEE (2016), pp. 193–193
12. M. Stonebraker, P. Brown, D. Zhang, J. Becla, Scidb: a database management system for applications with complex analytics. Comput. Sci. Eng. **15**(3), 54–62 (2013)
13. A. Bernasconi, A. Canakoglu, M. Masseroli, S. Ceri, *Meta-Base: A Novel Architecture for Large-Scale Genomic Metadata Integration* (IEEE/ACM Trans. Comput. Biol, Bioinf, 2020)
14. J.N. Weinstein, E.A. Collisson, G.B. Mills, K.R. Mills Shaw, B.A. Ozenberger, K. Ellrott, I. Shmulevich, C. Sander, J.M. Stuart, Cancer Genome Atlas Research Network, et al. The cancer genome atlas pan-cancer analysis project. Nature Genetics, **45**(10), 1113 (2013)
15. A. Kundaje, W. Meuleman, J. Ernst, M. Bilenky, A. Yen, A. Heravi-Moussavi, P. Kheradpour, Z. Zhang, J. Wang, M.lJ. Ziller et al. Integrative analysis of 111 reference human epigenomes. Nature, **518**(7539), 317 (2015)
16. ENCODE Project Consortium et al. The encode (encyclopedia of dna elements) project. Science **306**(5696), 636–640 (2004)
17. L. Nanni, P. Pinoli, A. Canakoglu, S. Ceri, Pygmql: scalable data extraction and analysis for heterogeneous genomic datasets. BMC Bioinformatics **20**(1), 560 (2019)
18. L. Nanni, P. Pinoli, A. Canakoglu, S. Ceri, Exploring genomic datasets: From batch to interactive and back, in *Proceedings of the 5th International Workshop on Exploratory Search in Databases and the Web*, ExploreDB 2018 (ACM, New York, NY, USA 2018), pp. 3:1–3:6
19. P. Pinoli, S. Ceri, D. Martinenghi, L. Nanni, Metadata management for scientific databases. Inf. Syst. **81**, 1–20 (2019)
20. L. Nanni, S. Ceri, C. Logie, Spatial patterns of ctcf sites define the anatomy of tads and their boundaries. Genome Biol. **21**(1), 1–25 (2020)
21. J.R. Dixon, D.U. Gorkin, B. Ren, Chromatin domains: the unit of chromosome organization. Molecular Cell **62**(5), 668–680 (2016)
22. M.H. Nichols, V.G. Corces, A ctcf code for 3d genome architecture. Cell **162**(4), 703–705 (2015)
23. S.S.P. Rao, M.H. Huntley, N.C. Durand, E.K. Stamenova, I.D. Bochkov, J.T. Robinson, A.L. Sanborn, I. Machol, A.D. Omer, E.S. Lander et al., A 3d map of the human genome at kilobase resolution reveals principles of chromatin looping. Cell **159**(7), 1665–1680 (2014)
24. S. Babaei, A. Mahfouz, M. Hulsman, B.P.F. Lelieveldt, J. de Ridder, M. Reinders, Hi-C chromatin interaction networks predict co-expression in the mouse cortex. PLoS Comput. Biol. **11**(5), e1004221 (2015)
25. O. Delaneau, M. Zazhytska, C. Borel, G. Giannuzzi, G. Rey, C. Howald, S. Kumar, H. Ongen, K. Popadin, D. Marbach et al. Chromatin three-dimensional interactions mediate genetic effects on gene expression. Science **364**(6439), eaat8266 (2019)
26. G. Kustatscher, P. Grabowski, J. Rappsilber, Pervasive coexpression of spatially proximal genes is buffered at the protein level. Molecular Syst. Biol. **13**(8), 937 (2017)
27. B. Zhang, S. Horvath, A general framework for weighted gene co-expression network analysis. *Statistical applications in genetics and molecular biology*, **4**(1) (2005)

28. D. Tian, R. Zhang, Y. Zhang, X. Zhu, J. Ma, MOCHI enables discovery of heterogeneous interactome modules in 3D nucleome (2019)
29. N. Zhou, I. Friedberg, M.S. Kaiser, Hierarchical markov random field model captures spatial dependency in gene expression, demonstrating regulation via the 3D genome. *bioRxiv*, page 2019.12.16.878371 (Dec 2019)
30. M. Varrone, L. Nanni, G. Ciriello, S. Ceri, Exploring chromatin conformation and gene co-expression through graph embedding. Bioinformatics, **36**(Supplement_2):i700–i708, 2020
31. Y. Bengio, A. Courville, P. Vincent, A review and new perspectives, Representation Learning (2012)
32. X. Yue, Z. Wang, J. Huang, S. Parthasarathy, S. Moosavinasab, Y. Huang, M.S. Lin, W. Zhang, P. Zhang, H. Sun, Graph embedding on biomedical networks: methods, applications, and evaluations. arXiv preprint arXiv:1906.05017 (2019)
33. A. Grover, J. Leskovec, node2vec: scalable feature learning for networks, in *Proceedings of the 22nd ACM SIGKDD International Conference on Knowledge Discovery and Data Mining*. ACM, pp. 855–864
34. L. Breiman, Random forests. Mach. Learn. **45**(1), 5–32 (2001)

Model, Integrate, Search... Repeat: A Sound Approach to Building Integrated Repositories of Genomic Data

Anna Bernasconi

Abstract A wealth of public data repositories is available to drive genomics and clinical research. However, there is no agreement among the various data formats and models; in the common practice, data sources are accessed one by one, learning their specific descriptions with tedious efforts. In this context, the integration of genomic data and of their describing metadata becomes—at the same time—an important, difficult, and well-recognized challenge. In this chapter, after overviewing the most important human genomic data players, we propose a conceptual model of metadata and an extended architecture for integrating datasets, retrieved from a variety of data sources, based upon a structured transformation process; we then describe a user-friendly search system providing access to the resulting consolidated repository, enriched by a multi-ontology knowledge base. Inspired by our work on genomic data integration, during the COVID-19 pandemic outbreak we successfully re-applied the previously proposed *model-build-search paradigm*, building on the analogies among the human and viral genomics domains. The availability of conceptual models, related databases, and search systems for both humans and viruses will provide important opportunities for research, especially if virus data will be connected to its host, provider of genomic and phenotype information.

1 Introduction

Genomics was born in relatively recent times: in the last two decades, after the introduction of Next Generation Sequencing (NGS) technologies [27], the processes of DNA/RNA sequencing have benefit form notable cost and time reductions. NGS data is employed at three levels. *Primary analysis* produces raw datasets of nucleotide bases, reaching a typical size of 200 Gigabytes per single human genome when stored [30]. *Secondary analysis* produces regions of interest, including *mutations*

A. Bernasconi (✉)

Dipartimento di Elettronica, Informazione e Bioingegneria, Politecnico di Milano, Via Ponzio 34/5, 20133 Milano, Italy

e-mail: anna.bernasconi@polimi.it

© The Author(s) 2022

L. Piroddi (ed.), *Special Topics in Information Technology*,

PoliMI SpringerBriefs, https://doi.org/10.1007/978-3-030-85918-3_8

(where the code of an individual differs from the code of the "reference" human being) – possibly associated with genetic diseases and cancers, *gene expression* (indicating in which conditions genes are active), and *epigenomic signals* (phenotype changes not involving alterations in the genetic sequence). Finally, *tertiary analysis* aggregates and combines together heterogeneous datasets produced during the preceding phase, trying to "making sense" of the data, unveiling complex biological mechanisms.

For boosting the last—and most interesting—type of analysis, thousands of datasets are becoming available every day, typically produced within the scope of large cooperative efforts, open for public use and made available for secondary research use [13], including the Encyclopedia of DNA Elements (ENCODE, [22]), Genomic Data Commons (GDC, [23]), Gene Expression Omnibus (GEO, [3]) Roadmap Epigenomics [24], and the 1000 Genomes Project [1]. In addition to these well-known sources, we are witnessing the birth of several initiatives of population-specific or nation-scale sequencing [29].

In the following, we focus on *processed genomic datasets*, which include *experimental observations*—representing regions along the genome chromosomes with their properties—and *metadata*, carrying information about the observed biological phenomena. The integration of genomic data and of their describing metadata is a challenge that is at the same time important (as a wealth of public data repositories is available to drive biological and clinical research), difficult (as the domain is complex and there is no agreement among the various data formats and definitions), and well-recognized (because, in the common practice, repositories are accessed one-by-one, with tedious and error-prone efforts). Although the potential collective amount of available information is huge, the effective combination of genomic datasets is hindered by their heterogeneity (in terms of download protocols, formats, notations, attribute names, and values) and lack of interconnectednes.

Motivating Example Let us consider a researcher who is looking for data to perform a comparison study between a human non-healthy breast tissue, affected by carcinoma, and a healthy sample coming from the same tissue type. Exploiting her previous experience, the researcher locates three portals having interesting data for this analysis (see Fig. 1). On GDC, one or more cases can be retrieved with the query "disease = Breast Invasive Carcinoma". To compare such data with references, the researcher chooses additional datasets coming from cell lines, a standard benchmark for investigations. On the GEO web interface, tumor cell line data is found by browsing thousands of human samples (e.g., the "T47D-MTVL" exhibits the disease "breast cancer ductal carcinoma"). Finally, on ENCODE, the researcher chooses both a tumor cell line ("MCF-7", affected by "Breast cancer (adenocarcinoma)") and a normal cell line("MCF-10A", widely considered the non-tumorigenic counterpart), to make a control comparison. As it can be noted, from both points of view of attributes and of values—when searching for disease-related information—we find many forms, only possibly pointing to comparable samples. This kind of information is not encoded in a unique way over data sources and is often missing. Considerable external knowledge is necessary to find appropriate connections; this cannot be obtained on the mentioned portals, but needs to be retrieved manually by querying specific databases, dedicated forums, or specialized ontologies.

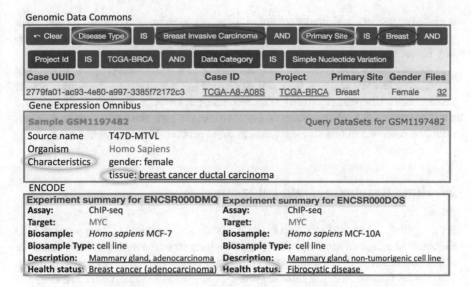

Fig. 1 Example of search results on GDC, GEO, and ENCODE. Yellow marks highlight how the 'disease' category is named in different sources; red marks show disease-related values

In this chapter we describe the research carried out in the context of the Genomic Computing ERC project [20], concerned with designing and building a repository of processed NGS data genomic datasets using a systematic and repeatable approach:

- *Model*: we analyze the domain state of the art (including scouting of online resources/documentation and testing data retrieval methods). Data is studied with the goal of proposing a conceptual model for the main characteristics shared by relevant data sources in the field, targeting completeness but favoring simplicity, for producing easy-to-use systems for biologists and genomic experts.
- *Integrate and build*: we select interesting open data sources for the domain, build solid pipelines to download data from them, and transform it into a standard interoperable format, obtaining a repository of homogenized data, to be used seamlessly from a unique endpoint, allowing integrative biological queries.
- *Search*: we target the end-users of the repository, i.e., experts of the domain who browse the repository in search for datasets to prove or disprove their research hypotheses. Interfaces need to take into account their background: considerable biological knowledge, but limited understanding of programming languages.

During the first phase of the COVID-19 epidemic, in March and April 2020, we responded proactively to the call to arms issued by the broad scientific community. We conducted an extensive requirement analysis by engaging in interdisciplinary conversations with a variety of scientists, including virologists, geneticists, biologists, and clinicians. This preliminary activity convinced us of the need for a structured proposal for viral data modeling and management. We thus reapplied the previously proposed methodology of modeling a data domain, integrating many sources to build a global repository, and finally making its content searchable to enable further analysis. This experience suggests that our approach is general enough to be applied to any domain of life sciences and encourages broader adoption.

Chapter organization. Section 2 describes our data integration proposal in the field of genomic tertiary analysis; Sect. 3 is dedicated to understanding the world of viral sequences and their descriptions (collected samples, host characteristics, variants and their impact on the related disease); Sect. 4 foresees how the two previous sections could be included in one single system to drive powerful biological discovery.

2   Human Genomic Data Integration

By analysing the players involved in the genomic data context [13], we proposed the Genomic Conceptual Model (GCM, [7]), which captures the most common metadata attributes that describe genomic data items and experiments in the available sources. The model (drawn in Fig. 2 represents a typical genomic region data file (i.e., the Item), using different perspectives: the biology (including information on the donor, the derived biological samples and different experimental replicates), the technology (with assay details and observed features), the management and organization aspects of the experiment, and the extraction parameters. GCM sets the basis for querying the underlying data sources for locating relevant experimental datasets.

We formalized a methodology for integrating and making datasets interoperable, called META-BASE architecture [4, 11], which is focused on obtaining a usable and consistent result, also from the data quality point of view [5]. As shown in Fig. 3, META-BASE takes care of retrieving datasets as partitions of a number of relevant data sources, transforming them into semi-structured datasets with an interoperable format (GDM [25]), cleaning redundant information, and joining their schemata into a global view represented by the GCM logical and physical implementation into a relational database. Later, we apply a value-normalization and enrichment procedure

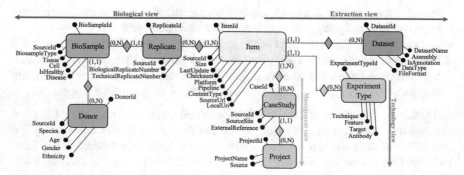

Fig. 2 The genomic conceptual model: the central entity ITEM is described along four perspectives.

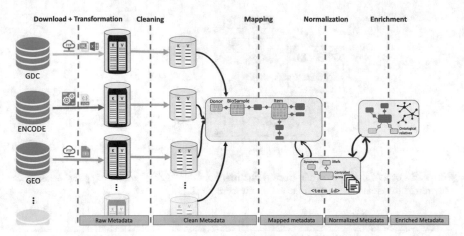

Fig. 3 META-BASE architecture, including six progressive phases

that exploits the content of several curated biomedical ontologies (such as OBI, NCIT, UBERON) for pointing to concepts (and their hierarchies) that are well-known in the domain [8]. The pipeline is general, open and extensible, being able to easily incorporate any number of new sources.

The resulting repository—already integrating several important sources such as ENCODE, The Cancer Genome Atlas from GCC [19] (via the OpenGDC framework), Roadmap Epigenomics, and the 1000 Genomes Project, contains now about 560K items, grouped within 67 datasets, collectively occupying more than 9TB of memory. A noteworthy metadata extraction framework has been implemented for the GEO source [18], where we employ a transformer-based machine learning approach to gather structured metadata from the experiments' textual descriptions.

The repository is exposed by means of user interfaces to respond to biological researchers' needs. We provide two different interfaces. The first one is a graph-based endpoint for expert users, who are focused on understanding the inference process performed on metadata in order match items in the repository [10]. Thanks to the ontological enrichment of metadata, we allow for a powerful semantic search mechanism. Consider, for example, choosing the tissue "Uterus". As shown in Fig. 4, this concept subsumes several other terms, which can be matched when different search levels are selected. In Table 1 we show the number of genomic items (i.e., region data files) that can be retrieved by using the `Orig.` level (only matching items described by exact keywords), `Syn.` level (also matching items described by synonyms and alternative forms), or `Exp.` level (also matching the sub-concepts of "uterus"), respectively resulting into the retrieval of 57, 1708, or 16851 items.

The second interface, GenoSurf [16] (http://gmql.eu/genosurf/) is a user-friendly search system providing access to the consolidated repository of metadata attributes, also enriched by a multi-ontology knowledge base, locating relevant genomic datasets, which can then be analyzed with off-the-shelf bioinformatic tools. The section of the Data Search interface of GenoSurf results from the translation of the

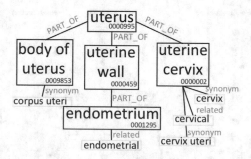

Fig. 4 Excerpt of the Uberon sub-tree originating from the "uterus" root. For space reasons, we only report the elements that are relevant to our example

Table 1 Items retrieved using different keywords from the "uterus" concept area.

Term ID	Search keyword	Orig.	Syn.	Exp.
0000995	Uterus	57	1708	16851
	uterus nos	1651	1708	16851
0009853	Body of uterus	0	9535	9535
	Corpus uteri	9535	9535	9535
0000002	Uterine cervix	0	5585	5585
	Cervix uteri	5417	5585	5585
	Cervix	167	5585	5585
	Cervical	1	5585	5585
0000459	Uterine wall	0	0	23
0001295	Endometrium	21	23	23
	Endometrial	2	23	23

GCM model into a much simpler denormalized structure consisting of a star with four related dimensions. This interface was evaluated by running an extended empirical study whose participants were knowledgeable in Biology and Computer Science. We collected many relevant insights related to the data scouting and extraction habits of different user profiles [9].

The frameworks and tools described in this section are included in a follow-up project, to be exploited for providing biologists and clinicians with a complete data extraction/analysis environment [21] that is: (i) guided by a conversational interface; (ii) equipped with a "marketplace" of ready-to-use best practices; we thus aim to break down the technological barriers that are currently hindering the practical adoption of our systems.

3   Virus Sequence Data Integration

Inspired by our work on genomic data integration, during the outbreak of the COVID-19 pandemic we searched for effective ways to help mitigate its effects with our contribution. As a first step, we conducted several interviews to experts and candidate users of our perspective systems [6], to quickly build the necessary expertise to operate in this new field. We understood that the collection of viral genome sequences is of paramount importance, in order to study the origin, wide spreading and evolution of SARS-CoV-2 (the virus responsible for the COVID-19 disease). Since the beginning of the pandemic, we have observed an almost exponential growth of the number of sequences deposited to known databases [14]: from few hundreds in March 2020, to one hundred thousand in August 2020, to almost two millions as of June 2021. Note that this is the first time that NGS technologies are been used for sequencing a massive amount of viral sequences. In several cases, also relevant associated data and metadata are provided, although their amount, coverage and harmonization are still limited.

Several institutions provide databases and resources for depositing viral sequences. Some of them, such as NCBI's GenBank [26], preexist the COVID-19 pandemic, as they host thousands of viral species – including, e.g., Ebola, SARS and Dengue. Other organizations have produced new data collections specifically dedicated to the hosting of SARS-CoV-2 sequences. It is the case of COG-UK [31], a pioneering project in the United Kingdom that has produced about one fourth of world-wide SARS-CoV-2 sequuences, and GISAID [28], which has soon become the worldwide predominant data source. While GenBank and COG-UK have adopted a fully open-source model of data distribution and sharing, GISAID is protecting the deposited sequences by having users login from an institutional site and accept a Database Access Agreement. We decided to re-apply the model-build-search paradigm used for human genomics. Building integrated databases for viral genomics and related search systems for accessing them is of uttermost importance for controlling the current pandemic and future ones. Several molecular biology studies can then be supported, considering haplotypes (i.e., clusters of inherited variations at single positions genomic sequence), phylogenetic trees (i.e., diagrams for representing the evolutionary relationships among organisms), and the evolution of new variants in general. Unfortunately, several limitations are still present, including the impact of GISAID's model, the lack of metadata quality, and the (un)willingness of sequence sharing, especially in some countries.

Understanding viruses from a conceptual modeling perspective is very important. Even if the domain of viral genomics is completely new, it presents many analogies with our previous challenges. Here, we model viral nucleotide sequences as strings of letters, with corresponding sub-sequences – the genes – that encode for amino acid proteins [12]. In our Viral Conceptual Model (Fig. 5) the Sequence of the virus is the central information. We describe a sequence's aspects using four perspectives: (i) technological (about the sequencing experiment); (ii) biological (about the virus—belonging to a complex taxonomy—isolated from an infected host and the isolation

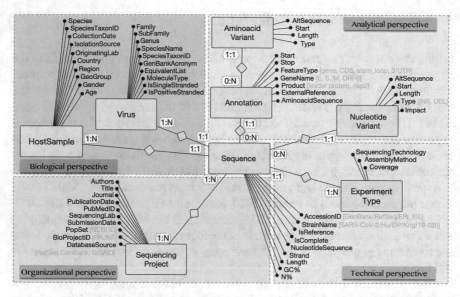

Fig. 5 The Viral Conceptual Model: the central fact SEQUENCE is described along four perspectives (biological, technical, organizational and analytical)

source from which viral material is extracted); (iii) organizational (the sequencing project, the hosting databases, the scientific and medical publications related to the discovery of sequences); (iv) analytical (the sequence's annotated parts—known genes, coding and untranslated regions—and the nucleotide/amino acids variants, computed with respect to the reference sequence of the species).

We then integrate sequences with their metadata from a variety of different sources. When stored into a unique database, virus strains may be searched and compared intra- and cross-species. We propose the powerful search interface ViruSurf [17] (http://gmql.eu/virusurf/), able to quickly extract sequences based on their combined variants, to compare different conditions, and to build interesting populations for downstream analysis. When applied to SARS-CoV-2, the virus responsible for COVID-19, complex conceptual queries upon our system are able to replicate the search results of recent articles, hence demonstrating considerable potential in supporting virology research. ViruSurf has being extended to accommodate other types of data, e.g., epitopes, which are sub-sequences of amino acids from a virus protein antigen that can activate an immune response from the host, being relevant for vaccine design. We produced EpiSurf (http://gmql.eu/episurf/, https://doi.org/10.1093/database/baab059), a web server for selecting viral populations of interest and analyzing how their amino acid changes are distributed along epitopes. In addition, we have provisioned the two search servers with the visual and analytical support by VirusViz [15] (http://gmql.eu/virusviz/); the application allows to show distributions of nucleotide and amino acid mutations, build groups (i.e., sequence populations of interest) and to compare their variation distributions. VirusViz provides examples related to SARS-CoV-2 variants of concern/interest (initially observed in UK, California, and New York), demonstrating how new variants can be traced since their starting dates (see https://github.com/DEIB-GECO/VirusViz/wiki/Supplementary-material-examples). VirusViz is also enriched by a knowledge base of amino acid changes effects (spanning from protein stability to epidemiological/immunological aspects), comprising for example viral transmission, fitness, binding affinity to host receptor and sensitivity to specific treatments (see CoV2K, [2]).

4 Conclusions

In Sect. 2, we presented our approach to modeling human genomic data, building a sound repository of genomic datasets using data integration techniques, and exposing its content through user interfaces that are rich in functionalities and data complexity. Our commitment is to continue the inclusion of relevant data sources for bioinformatics tertiary analysis, improving our process from a data quality and interoperability point of view. In Sect. 3, we have presented our approach to modeling viral sequences, building a repository collecting data from different viral data sources, and exposing its content over a first web interface, with enhanced functionalities on variant selection and filtering. This work has been realized in the first nine months of the SARS-CoV-2 epidemic worldwide. After setting the first milestones, we have already moved forward, considering the next challenges of this new domain with growing interest. More in general, the results on this thesis are part of a broad vision: availability of conceptual models, related databases and search systems for both human and viral genomics will provide important opportunities for genomic and clinical research, especially if virus (or other pathogens) sequences can be connected to the genotype and phenotype information regarding its host, i.e., the human organism, as suggested by Fig. 6. Such integration would drive more powerful biological discovery, being of particular interest for future epidemics events.

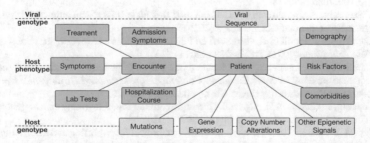

Fig. 6 Schema of patient phenotype for a viral disease linked to heterogeneous genomic information and to the sequence of the infecting virus, bridging Sects. 2 and 3 of this chapter

References

1. 1000 Genomes Project Consortium, A global reference for human genetic variation. Nature **526**(7571), 68 (2015)
2. R. Al Khalaf, T. Alfonsi et al., CoV2K: A Knowledge Base of SARS-CoV-2 Variant Impacts, in *Research Challenges in Information Science (RCIS 2021)* (Springer, Cham, 2021)
3. T. Barrett, S.E. Wilhite et al., NCBI GEO: archive for functional genomics data sets-update. Nucleic Acids Res. **41**(D1), D991–D995 (2012)
4. A. Bernasconi, Using metadata for locating genomic datasets on a global scale, in *Data and Text Mining in Biomedical Informatics (DTMBio 2018), CEUR Workshop Proceedings*, vol. 2482 (2018)
5. A. Bernasconi, Data quality-aware genomic data integration. Comput. Meth. Prog. Biomed. Update **1**, 100009 (2021)
6. A. Bernasconi, Extreme requirements elicitation: lessons learnt from the COVID-19 case study, in *Requirements Engineering: Foundation for Software Quality (REFSQ 2021), CEUR Workshop Proceedings*, vol. 2857 (2021)
7. A. Bernasconi, S. Ceri et al., Conceptual Modeling for Genomics: Building an Integrated Repository of Open Data, in *Conceptual Modeling (ER 2017)* (Springer, Cham, 2017), pp. 325–339
8. A. Bernasconi, A. Canakoglu et al., Ontology-driven metadata enrichment for genomic datasets, in *Semantic Web Applications and Tools for Life Sciences (SWAT4LS 2018), CEUR Workshop Proceedings*, vol. 2275 (2018)
9. A. Bernasconi, A. Canakoglu, S. Ceri, Exploiting conceptual modeling for searching genomic metadata: a quantitative and qualitative empirical study, in *Advances in Conceptual Modeling (EmpER 2019)* (Springer, Cham, 2019), pp. 83–94
10. A. Bernasconi, A. Canakoglu, S. Ceri, From a conceptual model to a knowledge graph for genomic datasets, in *Conceptual Modeling (ER 2019)* (Springer, Cham, 2019), pp. 352–360
11. A. Bernasconi, A. Canakoglu et al., META-BASE: a novel architecture for large-scale genomic metadata integration. IEEE/ACM Trans. Comput. Biol. Bioinf. (2020)
12. A. Bernasconi, A. Canakoglu et al., Empowering Virus Sequence Research through Conceptual Modeling, in *Conceptual Modeling (ER 2020)* (Springer, Cham, 2020), pp. 388–402
13. A. Bernasconi, A. Canakoglu et al., The road towards data integration in human genomics: players, steps and interactions. Briefings Bioinformat. **22**(1), 30–44 (2021)
14. A. Bernasconi, A. Canakoglu et al., A review on viral data sources and search systems for perspective mitigation of COVID-19. Briefings in Bioinformat. **22**(2), 664–675 (2021)
15. A. Bernasconi, A. Gulino et al., VirusViz: comparative analysis and effective visualization of viral nucleotide and aminoacid variants. Nucleic Acids Res. **49**(15), e90 (2021). https://doi.org/10.1093/nar/gkab478
16. A. Canakoglu, A. Bernasconi et al., GenoSurf: metadata driven semantic search system for integrated genomic datasets. Database (2019)
17. A. Canakoglu, P. Pinoli et al., ViruSurf: an integrated database to investigate viral sequences. Nucleic Acids Res **49**(D1), D817–D824 (2021)
18. G. Cannizzaro, M. Leone, et al., Automated integration of genomic metadata with sequence-to-sequence models. in *Joint European Conference on Machine Learning and Knowledge Discovery in Databases.* (Springer, Cham, 2020), pp. 187–203
19. E. Cappelli, F. Cumbo et al., OpenGDC: unifying, modeling, integrating cancer genomic data and Clinical Metadata. Appl. Sci. **10**(18), 6367 (2020)
20. S. Ceri, A. Bernasconi et al., Overview of GeCo: a project for exploring and integrating signals from the genome, in *Data Analytics and Management in Data Intensive Domains (DAM-DID/RCDL 2017)* (Springer, Cham, 2018), pp. 46–57
21. P. Covari, S. Pidò et al., GeCoAgent: a conversational agent for empowering genomic data extraction and analysis. in *ACM Transactions on Computing for Healthcare (HEALTH)* (2021)
22. C.A. Davis, B.C. Hitz et al., The Encyclopedia of DNA elements (ENCODE): data portal update. Nucleic acids Res. **46**(D1), D794–D801 (2018)

23. R.L. Grossman, A.P. Heath et al., Toward a shared vision for cancer genomic data. New England J. Med. **375**(12), 1109–1112 (2016)
24. A. Kundaje, W. Meuleman et al., Integrative analysis of 111 reference human epigenomes. Nature **518**(7539), 317–330 (2015)
25. M. Masseroli, A. Kaitoua et al., Modeling and interoperability of heterogeneous genomic big data for integrative processing and querying. Methods **111**, 3–11 (2016)
26. E.W. Sayers, M. Cavanaugh et al., GenBank. Nucleic Acids Res. **47**(D1), D94–D99 (2019)
27. S.C. Schuster, Next-generation sequencing transforms today's biology. Nature methods **5**(1), 16 (2007)
28. Y. Shu, J. McCauley, GISAID: Global initiative on sharing all influenza data–from vision to reality. Eurosurveillance **22**(13) (2017)
29. Z. Stark, L. Dolman et al., Integrating genomics into healthcare: a global responsibility. Am. J. Human. Genet. **104**(1), 13–20 (2019)
30. Z.D. Stephens, S.Y. Lee et al., Big Data: Astronomical or Genomical? PLOS Biol. **13**(7), 1–11 (2015)
31. The COVID-19 Genomics UK (COG-UK) consortium, An integrated national scale SARS-CoV-2 genomic surveillance network. The Lancet Microbe **1**(3), E99–E100 (2020)

Configurable Environments in Reinforcement Learning: An Overview

Alberto Maria Metelli

Abstract Reinforcement Learning (RL) has emerged as an effective approach to address a variety of complex control tasks. In a typical RL problem, an agent interacts with the environment by perceiving observations and performing actions, with the ultimate goal of maximizing the cumulative reward. In the traditional formulation, the environment is assumed to be a fixed entity that cannot be externally controlled. However, there exist several real-world scenarios in which the environment offers the opportunity to *configure* some of its parameters, with diverse effects on the agent's learning process. In this contribution, we provide an overview of the main aspects of environment configurability. We start by introducing the formalism of the Configurable Markov Decision Processes (Conf-MDPs) and we illustrate the solutions concepts. Then, we revise the algorithms for solving the learning problem in Conf-MDPs. Finally, we present two applications of Conf-MDPs: policy space identification and control frequency adaptation.

1 Introduction

Artificial Intelligence (AI) [32] and *Machine Learning* (ML) [23] are becoming terms widespread in our daily life. The progressive increase of the amount of available *data* and the evolution of the computing systems have enabled ML to become a powerful and effective *decision-making* tool. The traditional taxonomy of ML paradigms includes *supervised*, *unsupervised*, and *reinforcement learning*. Nowadays, the former two have reached an almost mature level of development, having also achieved marvelous results, especially in image classification [16], hand-written text recognition [29], and recommendation systems [1]. Instead, *Reinforcement Learning* (RL) [36] has only recently emerged, beyond the research field, as a valuable approach for several real-world applications. RL can be thought of as the most com-

A. M. Metelli (✉)
Dipartimento di Elettronica, Informazione e Bioingegneria, Politecnico di Milano, Via Ponzio 34/5, 20133 Milano, Italy
e-mail: albertomaria.metelli@polimi.it

© The Author(s) 2022
L. Piroddi (ed.), *Special Topics in Information Technology*,
PoliMI SpringerBriefs, https://doi.org/10.1007/978-3-030-85918-3_9

Fig. 1 Graphical
representation of the
interaction between an agent
and an environment in an
MDP

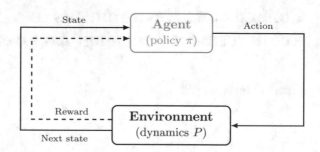

plete and general ML paradigm, to which supervised and unsupervised learning can
be reduced. Furthermore, it can be considered the paradigm closest to the intuitive
idea of the *learning* process, typical of biological entities.

The term *reinforcement* was introduced by Burrhus F. Skinner in behavioral psy-
chology to denote the attitude of organisms to strengthen a behavior if associated to
some desirable consequence [35]. Consider, for instance, a baby, an example of a
biological agent, learning how to walk. She will interact with the surrounding environment in a *trial and error* fashion, receiving positive and negative feedback. Pro-
gressively, she will learn the proper movements in order to stay upright and, finally,
effectively walk. In the context of AI, RL refers to the *computational* approach to
learning for artificial agents, in which the *interaction* between the agent and the
environment plays a central role.

The RL setting is composed of an (artificial) *agent* and an *environment* interacting
with one another [36]. The agent senses the state of the environment and performs
actions. Every action causes a transition of the environment to a new state, governed
by its dynamics (or transition model) P. The agent also receives from the environment a *reward* signal R. The agent-environment interaction proceeds in several
(possibly infinite) epochs. The ultimate goal of the agent consists in determining
a *policy* π, i.e., a prescription (possibly stochastic) telling which action to play in
every state. This form of interaction encodes a *sequential* decision-making problem,
typically modeled with the mathematical formalism of the *Markov Decision Processes* (MDPs) [30] (Fig. 1). The distinctive feature of RL, compared to the other
ML paradigms (supervised and unsupervised), is that the agent has to plan over a
possibly long horizon since the reward can be delayed. As a consequence, it might
be beneficial to sacrifice some reward achievable in the immediate future in order
to reach a more profitable region in the far future [36]. In the last decades, RL has
obtained remarkable success in several fields, including autonomous driving [13],
robotic locomotion [11], and video games [24], to mention a few.

Most of the RL literature considers the environment as a fixed entity, out of
any control. However, there exist several real-world scenarios in which a partial
intervention on the environment dynamics is allowed. Consider, for instance, the
task of learning how to drive a Formula 1 vehicle. The vehicle is a portion of the
environment and the driver has at her disposal several vehicle settings that can be
configured, while other parts of the environment are immutable. We call *environment
configuration* the activity of altering some environmental parameters.

In this contribution, we provide a summary of the Ph.D. dissertation entitled "Exploiting Environment Configurability in Reinforcement Learning" [18],[1] focused on studying the different aspects of environment configurability. The structure of the present contribution reflects the subdivision in parts of the dissertation. After having introduced, in Sect. 2, the idea of "environment configuration" and having provided some motivational examples, we outline the main results of the dissertation. In Sect. 3, we present the formalization of environment configurability, based on the novel Configurable Markov Decision Process (Conf-MDP) framework. Then, in Sect. 4, we briefly outline the approaches for learning in cooperative Conf-MDPs and focus on the experiment of car configuration based on TORCS. Finally, in Sect. 5, we present two applications of the Conf-MDP framework. We conclude, in Sect. 6, summarizing the results and discussing some future research directions.

2 Configurable Environments

As we mentioned in Sect. 1, the majority of the RL literature disregards the opportunity of configuring the environment, even when possible in the specific case of application. Indeed, traditionally, the modification of the environment dynamics during learning is considered the effect of a *non-stationary* process [2], i.e., a natural evolution of the environment. Instead, the possibility to *strategically* act on the environmental dynamics is studied in a limited number of works only. Some approaches belonging to the planning area [12, 38], some are constrained to specific forms of environment configurability [8, 9, 34], and others based on the *curriculum learning* framework [4, 7]. The goal of the dissertation [18] is to provide a uniform treatment of environment configurability in its diverse aspects. Before moving to the summary of the contributions, we present three motivational examples of environment configuration.

Example 1 (Car Configuration) Consider a Formula 1 driver that has to learn how to drive a Formula 1 car. The driver is the agent and the vehicle is a part of the environment, which is composed of other elements, like the road. The vehicle represents a configurable part of the environment since it is possible to change some of its settings (e.g., the wing orientation, the kind of tiers, and the brake repartition), whereas other parts of the environment are not configurable, such as psychical laws. The goals of the configuration may be different. First, we might want to find the vehicle settings that best fit the agent's needs, and allow she to learn the best performing policy. Second, we might want to train the agent with different vehicle configurations in order to speed up the learning process in a curriculum learning fashion. Notice that the vehicle configuration can be carried out by the agent itself, i.e., the driver can change some settings from its driving console, or by an external *configurator*, like a

---

[1] Online: http://hdl.handle.net/10589/170616.

Fig. 2 Graphical
representation of the
interaction between an agent
and an environment in a
Conf-MDP

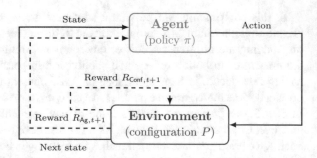

track engineer. Finally, both the agent and the external configurator share the same goals: improve the agent's learning experience, i.e., they act in what we denote as *cooperative* setting.

Example 2 (Teacher-Student) Another example of cooperative behavior is the interaction between a student and a teacher. They both aim at maximizing the knowledge acquired by the student, i.e., the agent. We can think of the teacher as either a physical person or an online teaching platform. In both cases, from the student's viewpoint, the *teaching style* is part of the environment and can be configured and should be tailored to the peculiarities of the student. Therefore, to select a suitable configuration, the teacher has to be aware of the student's capabilities that are to be inferred by interaction with the student herself.

Example 3 (Supermarket) We now consider an example in which agent and configurator no longer interact in a cooperative way. Suppose we are the owner of a supermarket and we have to configure the placement of the goods on the shelves in order to maximize our profit, so inducing customers to buy more. The customers, i.e., the agents, might have a different interest compared to that of the supermarket owner. Maybe they want to find the products they are interested in, in the smallest amount of time or buy certain goods only. This is an example in which the agents, the customers, and the configurator, the supermarket owner, show diverging interests. Thus, we are in what we denote as *non-cooperative* setting [31]. Moreover, we can distinguish between whether the agents are either aware or not of the configurator presence, leading to different levels of strategical behavior.

3 Modeling Environment Configurability

In this section, we outline the main contributions related to the modelization of configurable environments and the proposal of the corresponding solution concepts (Part I of [18]).

3.1 Configurable Markov Decision Processes

In order to represent the configuration opportunities the environment offers, we introduce an extension of the MDP framework: the *Configurable Markov Decision Process* (Conf-MDP) [22]. The main modification, compared to the traditional MDP, is that we no longer have a transition model P, governing the dynamics of the environment since environment configuration has the precise effect of changing it. Instead, we look at P just like the policy π, as an element that has to be determined as an output of the learning process. In the following, we will refer to P as environment *configuration*, instead of transition model, to highlight the features of the considered setting. Furthermore, to account for the possibly different interests of agent and configurator, we consider two reward functions R_{Ag} and R_{Conf} for agent and configurator, respectively (Fig. 2). With these two reward functions, we can define the performance indexes: the *expected returns* $J_{Ag}^{\pi,P}$ and $J_{Conf}^{\pi,P}$, i.e., the expected discounted sum of the rewards collected during the interaction with the environment:

$$J_{Ag}^{\pi,P} = \mathbb{E}_{Ag}^{\pi,P} \left[\sum_{t=0}^{\infty} \gamma^t R_{Ag,t+1} \right] \quad \text{and} \quad J_{Conf}^{\pi,P} = \mathbb{E}_{Conf}^{\pi,P} \left[\sum_{t=0}^{\infty} \gamma^t R_{Conf,t+1} \right],$$

where $\gamma \in [0, 1]$ is the discount factor that provides the relative importance between the reward collected in the present and those that will be collected in the future. The general goal in a Conf-MDP consists in finding an optimal policy π^* together with an optimal configuration P^*. These tasks are carried out by the agent and the configurator respectively. However, the notion of optimality strictly depends on the kind of interaction taking place between the agent and the configurator, as we discuss in the following section.

3.2 Solution Concepts

As we have illustrated in the examples presented in Sect. 2, the interaction between the agent and the configurator can take place in different forms. In particular, we distinguish between *cooperative* and *non-cooperative* Conf-MDPs.

In the cooperative setting, like in Examples 1 and 2, agent and configurator share the same objectives. In other words, they have the same reward function $R := R_{Ag} = R_{Conf}$. In such a case, defining a suitable solution concept is straightforward, as we look for an optimal policy π^* and an optimal configuration P^* they jointly maximize the expected return $J^{\pi,P}$, defined in terms of the unique reward function R:

$$(\pi^*, P^*) \in \arg\max_{(\pi,P)\in\Pi\times\mathcal{P}} \left\{ J^{\pi,P} \right\}. \tag{1}$$

It is worth noting that the search of the optimal policy π^* and the optimal configuration P^* is constrained in specific policy Π and configuration $\mathcal{P}$ spaces. The choice of these elements is highly relevant since it defines the actuations and the configuration possibilities. Indeed, while in a large body of the RL literature, it is assumed that the agent can play any policy (restriction are enforced for safety constraints in industrial applications [10]), it is in general unreasonable that the configurator can change arbitrarily the transition model. Indeed, in many real-world scenarios, the transition model groups portions of the environment that are immutable, like physical laws, and some that are mutable and, therefore, configurable (Example 1).

In the non-cooperative setting, like in Example 3, the agent and configurator reward functions R_{Ag} and R_{Conf} are kept distinct, to model situations in which the two entities have diverging interests. In this setting, in order to define suitable solution concepts, we have to resort to game-theoretic *equilibria* [33]. Moreover, the most suitable solution concept depends on the degree of *awareness* of the two entities about the presence of the other. While it is reasonable to assume that the configurator is always aware of the agent presence, the reverse might not be the case. The simplest situation is when the agent is unaware of the configurator presence. In such a case, it will react to any modification of the environment, perceived as a non-stationary evolution, with its optimal policy, that we call *best-response* policy, according to the game-theoretic terminology. Thus, we can map this setting to a *leader-follower* game in which the configurator (leader) wants to find the best configuration according to its reward function R_{Conf}, assuming that the agent (follower) will react with a best-response policy. The solution concept suitable for this setting is the *Stackelberg* equilibrium [37]:

$$ P^* \in \arg\max_{P \in \mathcal{P}} \left\{ J_{Conf}^{\beta_{Ag}(P), P} \right\}, $$

where $\beta_{Ag}(P) \in \arg\max_{\pi \in \Pi} \left\{ J_{Ag}^{\pi, P} \right\}$ is a best-response function that maps every configuration P to a an optimal policy $\beta_{Ag}(P) \in \Pi$ under P.[2] Differently, when the agent is aware of the configurator presence, we are in a more symmetric setting that can be mapped to a *simultaneous* game. In such a case, it is reasonable to consider the *Nash* equilibrium as a solution concept [26], in which we look for a policy-configuration pair such that neither the agent nor the configurator has an interest in unilaterally diverge from the equilibrium:

$$ \pi^* \in \arg\max_{\pi \in \Pi} \left\{ J_{Ag}^{\pi, P^*} \right\} \quad \text{and} \quad P^* \in \arg\max_{P \in \mathcal{P}} \left\{ J_{Conf}^{\pi^*, P} \right\}. $$

---

[2] The need for defining such a function derives from the fact that multiple best responses might generate different expected returns when evaluated with the configurator reward. In the game theory literature, common approaches consist in breaking ties in favor of the leader (strong Stackelberg equilibrium) or in favor of the follower (weak Stackelberg equilibrium) [5].

4 Learning in the Cooperative Configurable Markov Decision Processes

In this section, we provide a brief outline of the learning algorithms for cooperative Conf-MDPs, with particular reference to a car configuration example based on the TORCS simulator (Part II of [18]).

In the context of cooperative Conf-MDPs, the learning problem consists in finding an optimal policy π^* together with an optimal environment configuration P^* so that they jointly maximize the expected return, as in Equation (1). In [18], two algorithms are presented to tackle this problem. In the dissertation, we first focus on Conf-MDPs with finite state-action spaces and we propose a safe algorithm, *Safe Policy Model Iteration* (SPMI) [22], endowed with strong theoretical guarantees on the performance improvement. However, despite being the first attempt to solve the learning problem in Conf-MDPs, SPMI displays some limitations. First, it can be employed in finite Conf-MDPs only. Second, it requires the full knowledge of the environment dynamics, i.e., the configurator has to know not only the configurable parameters but also their effect on the transition dynamics. In order to overcome these limitations, we introduce a new algorithm, *Relative Entropy Model Policy Search* (REMPS) [19]. REMPS applies to continuous state-actions and no longer requires the knowledge of the transition model. The only assumption is that the configurator must know which are the configurable environmental parameters, while their effect on the transition model is learned from samples.

We tested REMPS on a simulated car configuration task based on the TORCS simulator [15]. The agent has access to a low-dimensional state representation based on the cars sensors (e.g., speed, focus, wheel speeds) and it can act on low-level controllers (acceleration, braking, and steering). The agent's goal consists of driving the car minimizing the lap time. The configurator, instead, is allowed to modify three configurable environmental parameters: rear wing orientation, front wing orientation, and front-rear brake repartition. In Fig. 3, we show the expected return and the average lap duration comparing REMPS, REPS (Relative Entropy Policy Search) [28], in

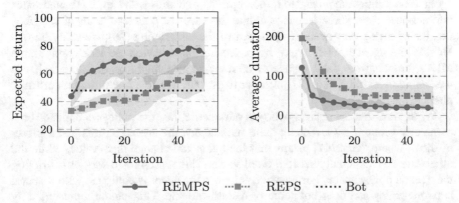

Fig. 3 Expected return and episode duration as a function of the number of iterations in the TORCS experiment (10 runs, 80% c.i.)

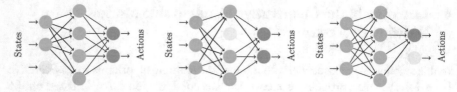

Fig. 4 An example of policy space modeled as a neural network with limitations in perceptions (left), mapping (center), and actuation (right)

which the car is not configured and the initial configuration is kept fixed, and a bot baseline. We can see that REMPS is able to outperform both REPS and the bot in terms of final policy performance and in terms of learning speed, showing that environment configuration can be beneficial also for speeding up the learning process.

5 Applications of Configurable Markov Decision Processes

In this section, we outline two cases of applications of Conf-MDPs in which the environment configuration opportunities can play a relevant role (Part III of [18]): *policy space identification* (Sect. 5.1) and *control frequency adaptation* (Sect. 5.2).

5.1 Policy Space Identification

In Example 2, about the teacher-student interaction, we have illustrated that in order to wisely choose the environment configuration, i.e., the teaching style, the configurator (teacher) has to be aware of the agent's (student) capabilities. More formally, the agent's capabilities are related to its perceptions, actuations, and ability to map states into actions. These three elements define the space of policies the agent can play, i.e., the *agent's policy space*. In this part of the dissertation [20], we study how to identify the agent's policy space by observing its behavior. Besides configurable MDPs, knowing the policy space of an expert agent might be of interest also in the imitation learning field [27] in order to prevent possible overfitting/underfitting phenomena.

We assume that the agent's policy is parametric, i.e., the policy π_θ depends on a parameter vector $\theta \in \Theta$. Among the θ parameters, the agent controls just a subset of them, where "controls" means that the agent can change their value, while the others are conventionally set to a fixed value. This represents a way of restricting the agent's policy space. For instance, suppose the agent is equipped with a neural network policy and does not perceive a state variable. This can be represented by setting to zero the weights related to that state variable. A similar construction can be performed to represent limitations in the agent's actuation and ability to map states to actions (Fig. 4).

In the dissertation [18], we propose an approach based on *generalized likelihood ratio* tests [3] to identify the set of parameters that the agent controls. Furthermore, we provide guarantees in terms of the probability of misidentification and numerical simulations on benchmark domains.

5.2 Control Frequency Adaptation

The typical RL setting deals with *discrete-time* problems that are obtained from the time discretization of a *continuous-time* problem [17]. Time discretization requires selecting a *control frequency* that is a design choice and represents, in all regards, configurable environmental parameter. This problem arises in several real-world domains, including robot control [14] and trading [25]. On one hand, we might be tempted to prefer high frequencies because they provide better control opportunities, leading to possibly more performing policies. However, in such a case, actions will last for a small time interval, thus, their effect on the environment will not be very clear. On the contrary, low frequencies sacrifice some control opportunities, but each action will last longer, making its effect more visible, with a possible benefit on the sample complexity. The question we address in this part of the dissertation [18] is whether we can exploit this trade-off to define a notion of optimal control frequency.

We propose to model the adaptation of the control frequency by means of *action persistence* [21], which consists in the repetition of each action for multiple k consecutive time steps. We focus on the trade-off in the choice of the persistence k. We first show that by increasing k, we give up control opportunities and, consequently, the optimal policy performance decreases. To visualize the beneficial effect of a large k on the sample complexity, we propose a novel algorithm, *Persistent Fitted Q-Iterations* (PFQI) [21]. PFQI is a batch RL algorithm and extends the classical Fitted Q-Iterations (FQI) [6] to account for action persistence. PFQI enjoys a sample complexity that decreases with the persistence k. Therefore, we observe that the optimal value of k depends on the number of samples (batch size) available for learning. Intuitively, when the batch size is large, we can afford a small value of k, whereas with few samples, we benefit from the regularization effect of using a high persistence. Furthermore, we propose a heuristic approach to suggest an approximately optimal value of k.

We tested PFQI on a simple forex trading simulator. In Fig. 5, we show the performance of the policy learned with PFQI for different batch sizes and different persistence values. First of all, we notice a generally improving trend as the batch size increases. If we look at small batch sizes, we observe that the best performance is obtained with high values of persistence ($k = 4$ or $k = 8$), whereas as the batch size increases, the best performance is attained by small values of persistence, namely $k = 1$ for batch size 400.

A. M. Metelli

Fig. 5 Expected return as a
function of the batch size in
the forex experiment (10
runs, 95% c.i.)

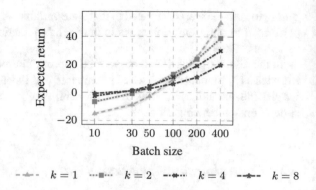

6 Conclusions

In this contribution, we summarized the results of the Ph.D. dissertation [18]. We
started by introducing the novel framework of the Conf-MDPs to model several
real-world situations in which agent interacts with a configurator, entitled to modifying some environmental parameters. Then, we provided learning algorithms for
the cooperative setting, showing that environment configuration can be beneficial
for the performance of the agent's policy. We have seen that knowing the agent's
policy space is important for suitably choosing an environment configuration and
we presented an approach for identifying it from samples. Finally, we have studied the effects on the learning performance of a specific configurable environmental
parameter, namely the control frequency.

We hope, with this dissertation, to have shed light on a novel topic of interest
in the RL community. Numerous directions could be further explored. From the
modelization standpoint, it would be worth considering the possibility of having
multiple agents interacting with multiple configutators acting in the same environment. Concerning the learning problem, online approaches for searching for the best
configuration in the cooperative setting should be investigated, while the learning
problem for the non-cooperative setting has to be further deepened. Finally, concerning action persistence, our approach is limited to a fixed persistence. It would be
interesting studying the possibility of having a control frequency that dynamically
changes during the learning process.

Acknowledgements The author wishes to thank Prof. Marcello Restelli for the continuous support
and guidance.

References

1. J. Bobadilla, F. Ortega, A. Hernando, A. Gutiérrez, Recommender systems survey. Knowl. Based Syst. **46**, 109–132 (2013)
2. B.L. Bowerman, *Nonstationary Markov Decision Processes and Related Topics in Nonstationary Markov Chains* (1974)
3. G. Casella, R.L. Berger, *Statistical Inference*, vol. 2 (Duxbury Pacific Grove, CA, 2002)
4. K.A. Ciosek, S. Whiteson, OFFER: off-environment reinforcement learning, in *Proceedings of the Thirty-First AAAI Conference on Artificial Intelligence, February 4-9, 2017, San Francisco, California, USA*, ed. by S.P. Singh, S. Markovitch (AAAI Press, 2017), pp. 1819–1825
5. V. Conitzer, T. Sandholm, Computing the optimal strategy to commit to, in *Proceedings 7th ACM Conference on Electronic Commerce (EC-2006), Ann Arbor, Michigan, USA, June 11-15, 2006*, ed. by J. Feigenbaum, J.C.-I. Chuang, D.M. Pennock (ACM, 2006), pp. 82–90
6. D. Ernst, P. Geurts, L. Wehenkel, Tree-based batch mode reinforcement learning. J. Mach. Learn. Res. **6**, 503–556 (2005)
7. C. Florensa, D. Held, M. Wulfmeier, M. Zhang, P. Abbeel, Reverse curriculum generation for reinforcement learning, in *1st Annual Conference on Robot Learning, CoRL 2017, Mountain View, California, USA, November 13-15, 2017, Proceedings*, vol. 78 of *Proceedings of Machine Learning Research* (PMLR, 2017), pp. 482–495
8. V. Gallego, R. Naveiro, D.R. Insua, Reinforcement learning under threats, in *The Thirty-Third AAAI Conference on Artificial Intelligence, AAAI 2019, The Thirty-First Innovative Applications of Artificial Intelligence Conference, IAAI 2019, The Ninth AAAI Symposium on Educational Advances in Artificial Intelligence, EAAI 2019, Honolulu, Hawaii, USA, January 27 - February 1, 2019* (AAAI Press, 2019), pp. 9939–9940
9. V. Gallego, R. Naveiro, D.R. Insua, D. Gómez-Ullate, Opponent aware reinforcement learning, in *CoRR*, abs/1908.08773 (2019)
10. J. García, F. Fernández, A comprehensive survey on safe reinforcement learning. J. Mach. Learn. Res. **16**, 1437–1480 (2015)
11. T. Haarnoja, S. Ha, A. Zhou, J. Tan, G. Tucker, S. Levine, Learning to walk via deep reinforcement learning, in *Robotics: Science and Systems XV, University of Freiburg, Freiburg im Breisgau, Germany, June 22-26, 2019*, ed. by A. Bicchi, H. Kress-Gazit, S. Hutchinson (2019)
12. S. Keren, L.E. Pineda, A. Gal, E. Karpas, S. Zilberstein, Equi-reward utility maximizing design in stochastic environments, in *Proceedings of the Twenty-Sixth International Joint Conference on Artificial Intelligence, IJCAI 2017, Melbourne, Australia, August 19-25, 2017*, ed. by C. Sierra, ijcai.org (2017), pp. 4353–4360
13. B. Ravi Kiran, I. Sobh, V. Talpaert, P. Mannion, A.A. Al Sallab, S.K. Yogamani, P. Pérez, Deep reinforcement learning for autonomous driving: a survey, in *CoRR*. abs/2002.00444 (2020)
14. J. Kober, J. Andrew Bagnell, J. Peters, Reinforcement learning in robotics: a survey. I. J. Robotics Res. **32**(11), 1238–1274 (2013)
15. D. Loiacono, A. Prete, P.L. Lanzi, L. Cardamone, Learning to overtake in TORCS using simple reinforcement learning, in *Proceedings of the IEEE Congress on Evolutionary Computation, CEC 2010, Barcelona, Spain, 18-23 July 2010* (IEEE, 2010), pp. 1–8
16. D. Lu, Q. Weng, A survey of image classification methods and techniques for improving classification performance. Int. J. Remote Sensing **28**(5), 823–870 (2007)
17. D.G. Luenberger, Introduction to dynamic systems: theory, models, and applications. Technical report (1979)
18. A.M. Metelli, *Exploiting Environment Configurability in Reinforcement Learning*. PhD thesis, Politecnico di Milano, March 2021
19. A.M. Metelli, E. Ghelfi, M. Restelli, Reinforcement learning in configurable continuous environments, in *Proceedings of the 36th International Conference on Machine Learning, ICML 2019, 9-15 June 2019, Long Beach, California, USA*, ed. by K. Chaudhuri, R. Salakhutdinov, vol. 97 of *Proceedings of Machine Learning Research* (PMLR, 2019), pp. 4546–4555
20. A.M. Metelli, G. Manneschi, M. Restelli, Policy space identification in configurable environments, in *CoRR*, abs/1909.03984 (2019)

21. A.M. Metelli, F. Mazzolini, L. Bisi, L. Sabbioni, M. Restelli, Control frequency adaptation via action persistence in batch reinforcement learning, in *Proceedings of the 37th International Conference on Machine Learning, ICML 2020, 13-18 July 2020, Virtual Event*, vol. 119 of *Proceedings of Machine Learning Research* (PMLR, 2020), pp. 6862–6873

22. A.M. Metelli, M. Mutti, M. Restelli, Configurable markov decision processes, in *Proceedings of the 35th International Conference on Machine Learning, ICML 2018, Stockholmsmässan, Stockholm, Sweden, July 10-15, 2018*, ed. by J.G. Dy A. Krause, vol. 80 of *Proceedings of Machine Learning Research* (PMLR, 2018), 3488–3497

23. T.M. Mitchell. *Machine Learning*, International Edition. McGraw-Hill Series in Computer Science (McGraw-Hill, 1997)

24. V. Mnih, K. Kavukcuoglu, D. Silver, A. Graves, I. Antonoglou, D. Wierstra, M.A. Riedmiller, Playing atari with deep reinforcement learning, in *CoRR*, abs/1312.5602 (2013)

25. J.E. Moody, M. Saffell, Learning to trade via direct reinforcement. IEEE Trans. Neural Netw. **12**(4), 875–889 (2001)

26. J. Nash, Non-cooperative games. Ann. Math. **54**(2), 286–295 (1951)

27. T. Osa, J. Pajarinen, G. Neumann, J. Andrew Bagnell, P. Abbeel, J. Peters, An algorithmic perspective on imitation learning. Foundations Trends Robot. **7**(1–2), 1–179 (2018)

28. J. Peters, K. Mülling, Y. Altun, Relative entropy policy search, in *Proceedings of the Twenty-Fourth AAAI Conference on Artificial Intelligence, AAAI 2010, Atlanta, Georgia, USA, July 11-15, 2010*, ed. by M. Fox, D. Poole (AAAI Press, 2010)

29. J. Puigcerver, Are multidimensional recurrent layers really necessary for handwritten text recognition?, in *2017 14th IAPR International Conference on Document Analysis and Recognition (ICDAR)*, vol. 1, pp. 67–72 (IEEE, 2017)

30. M.L. Puterman, *Markov Decision Processes: Discrete Stochastic Dynamic Programming*. John Wiley & Sons (2014)

31. G. Ramponi, A.M. Metelli, A. Concetti, M. Restelli, Online learning in non-cooperative configurable Markov decision process, in *AAAI-21 Workshop on Reinforcement Learning in Games* (2021)

32. S.J. Russell, P. Norvig, *Artificial Intelligence-- Modern Approach* (Third International Edition, Pearson Education, 2010)

33. L.S Shapley, Stochastic games. Proc. Natl. Acad. Sci. **39**(10), 1095–1100 (1953)

34. R. Silva, F.S. Melo, M. Veloso, What if the world were different? gradient-based exploration for new optimal policies, in *GCAI-2018, 4th Global Conference on Artificial Intelligence, Luxembourg, September 18-21, 2018*, ed. by D.D. Lee, A. Steen, T. Walsh, vol. 55 of *EPiC Series in Computing* (EasyChair, 2018), pp. 229–242

35. B.F. Skinner, *The Behavior of Organisms: An Experimental Analysis* (1938)

36. R.S. Sutton, A.G. Barto, *Reinforcement Learning: An Introduction*. MIT press (2018)

37. H. Von Stackelberg, *Marktform und gleichgewicht*. J. Springer (1934)

38. H. Zhang, Y. Chen, D.C. Parkes, A general approach to environment design with one agent, in *IJCAI 2009, Proceedings of the 21st International Joint Conference on Artificial Intelligence, Pasadena, California, USA, July 11-17, 2009*, ed. by C. Boutilier (2009), pp. 2002–2014

Machine Learning for Scientific Data Analysis

Gabriele Scalia

Abstract Over the last few years, machine learning has revolutionized countless areas and fields. Nowadays, AI bears promise for analyzing, extracting knowledge, and driving discovery across many scientific domains such as chemistry, biology, and genomics. However, the specific challenges posed by scientific data demand to adapt machine learning techniques to new requirements. We investigate machine learning-driven scientific data analysis, focusing on a set of key requirements. These include the management of uncertainty for complex data and models, the estimation of system properties starting from low-volume and imprecise collected data, the support to scientific model development through large-scale analysis of experimental data, and the machine learning-driven integration of complementary experimental technologies.

1 Introduction

Managing scientific data poses unique and challenging problems. In particular, the volume, complexity and heterogeneity of the available data has led to an increasing relevance of data-driven methodologies to integrate, process and analyze it. To match the ambitious application goals, the complex and high-dimensional data sources and the quality issues characterizing this data, increasingly sophisticated methods have been proposed, with a central role played by artificial intelligence (AI). Machine learning (ML) and deep learning-based methods are revolutionizing applications which have been consolidated for years [3, 16, 17] or enabling totally new directions in tandem with new technological progresses [7].

This combination, even tough extremely promising, is not without challenges, and requires an adaptation process on both sides [16]. On the one side, progresses in AI have lead to rethinking consolidated fields such as drug discovery [3] and medicinal chemistry synthesis [17], just to name a few. On the other, the specificities

G. Scalia (✉)
Dipartimento di Elettronica, Informazione e Bioingegneria, Politecnico di Milano, Via Ponzio 34/5, 20133 Milano, Italy
e-mail: gabriele.scalia@polimi.it

© The Author(s) 2022

L. Piroddi (ed.), *Special Topics in Information Technology*,
PoliMI SpringerBriefs, https://doi.org/10.1007/978-3-030-85918-3_10

of scientific data and individual fields demand the development of new methods or the adaptation of existing ones.

This work goes in the direction of exploring ML techniques for scientific data analysis, investigating how they can be tailored to the needs of the data in this context. In doing this, it focuses on a set of case studies spanning chemistry, biology and genomics. The common thread of this research is the set of challenges posed by scientific data, including *quality* limitations, experimental *uncertainty*, low *volume*, and *complementarity* of the available sources in terms of modalities, resolutions or measured ranges. Additionally, while complete and structured descriptions can exist in some cases, data ontologies are often only partial, evolving and semi-structured. In view of this, it appears relevant to investigate ML-driven methods and systems specifically tailored to scientific data management and analysis.

This is an interdisciplinary research that, focusing on the challenges emerging in fields such as chemistry, genomics and biomedical research, investigates ML-driven techniques to face a set of identified requirements: (1) the management of uncertainty for complex data and models such as deep neural networks (DNNs), (2) the estimation of system properties starting from imprecise, low-volume and evolving data, (3) the continuous validation of scientific models through large-scale comparisons with experimental data and (4) the unsupervised integration of multiple heterogeneous data sources related to different technologies to overcome individual technological limitations. Common to virtually all fields driven by experimental data, these requirements are faced through a set of case studies.

This chapter presents the main results included in the Ph.D. thesis [12]. In the following, each section focuses on one of the directions previously mentioned. Each direction is investigated in the context of a specific scientific application scenario, but results are general and can be easily extended to other domains.

2 Uncertainty Estimation and Domain of Applicability for Neural Networks

Experimental data are always characterized by some level of intrinsic variability and imprecision. Models trained on those data inherit that uncertainty and are also affected by another kind of uncertainty that comes from insufficient training samples, often difficult to quantify, especially for complex models such as DNNs. For these reasons, modeling uncertainty in DNNs has recently attracted great interest and currently represents a major research direction in the field [5]. Accounting for uncertainty does not only mean outputting a confidence score for a given input. It also means *changing the way predictions are made*, taking into account the concept of "unknown" during the training and/or inference phases, and *making sense of the resulting uncertainty estimates*, which need to reflect and satisfy some principles to be really useful and trustworthy for the users. The latter point is especially important since it has been shown that modern neural networks, even though really accurate, are often "over-confident" in their output probabilities (e.g., a DNN could say that a certain image is 99% likely to be a cat, while the true observed confidence is much lower) [5]. Recent progresses on uncertainty estimation in DNN emerged in the computer vision field, mainly focusing on convolutional neural networks (CNNs). In that context, uncertainty prediction is often studied to overcome interpretability and safety limitations of modern computer vision applications such as autonomous driving [6].

When we consider scientific data and applications, uncertainty estimation assumes a unique relevance. Experimental datasets are often comparatively small (being costly to generate), sparse, and affected by various kinds of inherent imprecision such as experimental errors, lack of coherent ontologies, and misreporting [16]. On top of this, common requirements of scientific applications put particular emphasis on uncertainty. For example, *drug discovery* is strictly related to exploring the "uncharted" chemical space and, therefore, estimating the uncertainty over such predictions is crucial since there will always be a *knowledge boundary* beyond which predictions start to degrade. In such cases, uncertainty estimation becomes strictly related to the problem of defining a *domain of applicability* for a model [13].

We study this problem in the context of *molecular property prediction*, formally referred to as *Quantitative Structure-Activity Relationship* (QSAR). In the last few years, pioneering neural network architectures for QSAR, such as graph neural networks (GNNs), have been proposed. Such models, combined to an increasing availability of data and computational power, have led to state-of-the-art performance for this task. However, these models are still characterized by some key limitations, such as interpretability and generalization ability [16].

In this respect, we investigate how uncertainty can be modeled in DNNs, theoretically reviewing existing methods and experimentally testing them on GNNs for molecular property prediction. In parallel, we develop a framework to qualitatively and quantitatively evaluate the estimated uncertainties from multiple points of view. An overview of the methodology is shown in Fig. 1.

2.1 A Bayesian Graph Neural Network for Molecular Property Prediction

Uncertainty can be the result of inherent data noise or could be related to what the model does not yet know. These two kind of uncertainties—*aleatoric* and *epistemic*—can be combined to obtain the total predictive uncertainty of the model. We extend a GNN to model both uncertainty components.

When not explicitly modeled, the inherent observation noise is assumed constant for every observed molecule. However, this assumption does not hold in many realistic settings, where input-dependent noise needs to be modeled, such as chemistry applications. Data-dependent aleatoric uncertainty is referred to as *heteroscedastic* and its importance for DNNs has been recently highlighted [6]. Since aleatoric uncer-

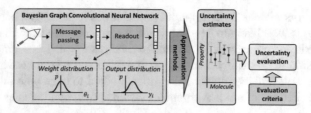

Fig. 1 Overview of the methodology described in [13]. Reprinted with permission from *G. Scalia, C. Grambow, B. Pernici, Y-P. Li, W. H. Green, J. Chem. Inf. Model. 2020, 60, 6, 2697–2717.* Copyright 2020 American Chemical Society

tainty is a property of data, it can be learned directly from the data adapting the model and the loss function. However, aleatoric uncertainty does not account for epistemic uncertainty. This can be overcome by performing Bayesian inference, through the definition of a *Bayesian neural network*.

In a Bayesian neural network the weights of the model θ are *distributions* learned from training data $\mathcal{D}$, instead of point estimates, and therefore it is possible to predict the output distribution **y** of some new input **x** through the *predictive posterior distribution* $p(\mathbf{y} \mid \mathbf{x}, \mathcal{D}) = \int p(\mathbf{y} \mid \mathbf{x}, \theta) p(\theta \mid \mathcal{D}) d\theta$. Monte Carlo integration over M samples of the posterior distribution can approximate the intractable integral, however, obtaining samples from the true posterior is virtually impossible for DNNs. Therefore, an approximate posterior $q(\theta) \approx p(\theta \mid \mathcal{D})$ is introduced. A common technique to derive $q(\theta)$ is *variational inference* (VI). Approximate VI can scale up to the training-intensive large datasets/models of modern applications. Major examples of techniques of this type are MC-Dropout and ensembling-based methods. In [13], different techniques have been experimentally compared.

Experimental results on major public datasets [13] show that the computed uncertainty estimates allow correctly approximating the expected errors in many cases, and, in particular, when test molecules are comparatively similar with respect to training molecules. When this is not the case, uncertainty tends to be underestimated, but it still allows ranking test predictions by confidence. Moreover, experiments show how modeling both types of uncertainty is in general beneficial, and that the relative contribution of each uncertainty type to total uncertainty is dataset dependent. Additionally, it has been shown how modeling uncertainty has a consistent positive impact on model's accuracy.

The methodology, experimental results and additional analyses are detailed in [13] and in Chap. 3 of [12].

3 Machine Learning Estimation of System Properties from Uncertain, Low-Volume Data

ML methods are commonly used to learn a model of the data (in terms of inputs or inputs/outputs). Generally speaking, every ML model can be thought as a model of the observed data, and, therefore, as an estimator of a function of the data distributions [4]. This definition is general and includes both supervised and unsupervised tasks.

However, in many scientific settings the ultimate goal is not learning a model of the data, but using the data to learn something about the underlying system that has generated it. The way scientific data and ML techniques can be used in such contexts necessarily changes and is generally less investigated. This direction is characteristic of scientific domains, and it is also hindered by all the various types of quality limitations characterizing data in this context, such as imprecision and scarcity [9].

We explore the problem of estimating some properties of a system (e.g., a biological system) starting from imprecise, low-volume data. As a case study, we consider *biological systems* characterized by some kind of internal information exchanges, and we develop a ML-driven methodology to estimate the optimal way of transferring information given as input a set in-silico/in-vitro experiments for the system.

3.1 A Machine Learning-Driven Approach to Optimize Bounds on the Capacity of a Molecular Channel

We consider biological systems characterized by some kind of internal information exchange. Such exchange can be described from a communication point of view, allowing defining a *capacity* for the communication channel (which, in this case, is a molecular channel). We face the problem of estimating this capacity, which is strictly related to estimating the optimal way of transferring information in the system, only using a set of inputs/outputs for the system, that can be, for example, the result of in-silico/in-vitro experiments. This is particularly useful since, for many biological systems, an analytical or statistical description does not exist, but datasets of inputs/outputs can be easily obtained.

We propose a novel methodology that frames the estimation of the capacity as the optimization problem of finding an upper and a lower bound on the true value. The bounds are optimized starting from the data and using an evolutionary iterative algorithm. Being estimated from the data, the accuracy of the resulting interval is affected by the uncertainty and the volume of the available data. Therefore, particular emphasis has been placed on overcoming data scarcity and managing uncertainty in the available biological data. On overview of the methodology is shown in Fig. 2.

Two fundamental factors hinder the optimization procedure, and are addressed by the proposed methodology:

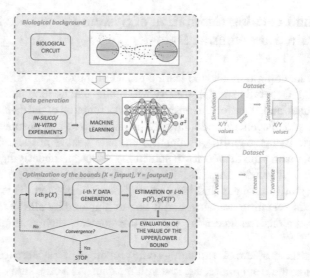

Fig. 2 Overview of the methodology presented in [10]. Using an iterative evolutionary optimization algorithm and a DNN-driven augmentation strategy, the proposed methodology converges to tight capacity bounds. © 2021 IEEE. Reprinted, with permission, from *F. Ratti, G. Scalia, B. Pernici and M. Magarini, "A Data-driven Approach to Optimize Bounds on the Capacity of the Molecular Channel", GLOBECOM 2020 - 2020 IEEE Global Communications Conference, Taipei, Taiwan, 2020, pp. 1–7*

- The biological model is a *black-box*: an analytical characterization of its behavior is not known, and therefore an analytical formulation of the bounds does not exist. This prevents the usage of exact or gradient-based optimization algorithms. We overcome this issue with two complementary solutions: (1) evaluating the bounds functions directly on the data, and (2) using a derivative-free optimization algorithm (the Covariance Matrix Adaptation Evolution Strategy (CMA-ES)).
- Given the fact that biological simulations are heavily time consuming, we cannot rely on on-demand simulations to approximate any possible input distribution during the execution of iterative optimization algorithm. We address this issue by using a set of previously generated simulations as input of a DNN-based *data augmentation* module.

The proposed methodology is experimentally evaluated on a in-silico system composed of two prokaryotic cells. The methodology and the experimental results are detailed in [10] and in Chap. 4 of [12].

4 Data-Driven Validation and Development of Scientific Models

In many scientific settings, models are developed independently and externally with respect to the collected data and the empirical observations (for example, in principle-based models). These models are not designed directly from the data, but experimental data still have a key role in the development/validation/refinement cycle. Necessarily, the role of ML methods in this context changes. Indeed, in this context they primarily take a *validation* role, supporting the development cycle of the model leveraging the available data.

The sharing of scientific data has greatly increased in the last decade, leading to open repositories in many different scientific domains, also thanks to specific initiatives and guidelines [18]. However, it was recently discussed how data sharing practices have received far more study than has data reuse, and that *data sharing* and *open data* are not final goals in themselves, but the real benefit is in *data reuse*, which is "an understudied problem that requires much more attention if scientific investments are to be leveraged effectively" [8]. However, in order to reuse datasets from multiple sources, a series of challenges must be addressed.

We discuss the design of a framework to manage the development, validation and refinement cycle of scientific models taking advantage of large amounts of scientific data (experiments) extracted from the literature. As a case study we consider chemical kinetics models, which determine the reactivity of fuels and mixtures, but the approach taken is general and domain-agnostic.

4.1 Towards an Integrated Framework to Support Scientific Model Development

We investigate this direction from an information systems point of view, discussing the requirements and an architecture for the framework. Moreover, we develop a prototype to better analyze the framework's requirements, that relies on a service-oriented architecture and provides a set of functions to import experiments and models, automatically run simulations and compute global validation indices/statistics.

The design of an integrated framework (see Fig. 3) to support model development through large-scale automatic validation on published experimental data requires addressing a set of use case [14, 15]. Those include: (1) *the acquisition of new experiments*, guaranteeing consistency, uniqueness and quality, (2) *the simulation of experiments through available models*, that requires automatically interpreting and handling data and models, (3) *cross comparisons and global validations*, including model/experiment validation through large-scale comparison, also supported by ML techniques, (4) *managing changes in models*, tracking and supporting the development through a continuous validation approach.

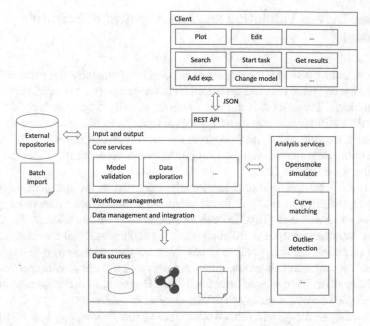

Fig. 3 Overview of the architecture of the SciExpeM framework, highlighting the main logical components. Reprinted from *G. Scalia, M. Pelucchi, A. Stagni, A. Cuoci, T. Faravelli, B. Pernici, Towards a scientific data framework to support scientific model development, Data Science, 2 (1–2), IOS Press, 2020* with permission from IOS Press. The publication is available at IOS Press through http://dx.doi.org/10.3233/DS-190017

The SciExpeM framework [14] has been designed to support these use cases, automatizing the processes of acquiring, interpreting, simulating and cross-comparing models and experiments. SciExpeM provides *model validation* functionalities supported by large scale *data aggregation* and *data analysis* tools. Analysis tools include outlier detection, clustering and statistical correlations. Examples of outcomes that can be achieved through SciExpeM include: (1) finding potential errors in the experimental data through cross-comparisons with other experiments/models (also identifying possible causes for such errors), (2) tracking models, highlighting their performance over different versions with respect to a set of experiments and identifying critical areas, (3) identifying models/experiments remarkably different with respect to others available in the system to be validated.

The SciExpeM framework has been first studied in [15], and then largely extended in [14]. Chapter 5 of [12] details the main results.

5 Unsupervised Deep Learning-Driven Integration of Multiple Sources

Data integration has been a central research direction across many fields and from different perspectives for decades, with applications in countless areas. Nonetheless, integration in the context of scientific data is still considered a fundamental challenge to be addressed. For example, the "integration of single-cell data across samples, experiments, and types of measurement" has been very recently highlighted as one of the main challenges in single-cell data science [7].

One limitation of traditional integration techniques is that of being mostly rule-driven. Instead, data-driven (including ML-driven) methodologies can address the lack of ontologies and complete descriptions of the underlying phenomena, not requiring strong a priori assumptions about the available data sources (in terms of quality, available modalities, etc.). Moreover, they can help exploit *complementary strengths* of the available data sources and enable context-aware analysis.

In the scientific domain, data integration is often an *enabler* for other activities. Integration can help overcoming data scarcity and quality issues, thus indirectly improving downstream data-driven applications which rely on large, high-quality datasets [16]. Integration can also allow overcoming technological limitations hindering next-generation applications. For example, creating comprehensive and multiscale biological atlases at single cell resolution of the human body is currently recognized as next frontier to understand cellular basis of health and disease [11]. However, a single experimental technology for doing that currently does not exist: multiple technologies capturing different aspects, scales and modalities are available.

We investigate this problem designing a novel methodology, named *Tangram* [1], to (1) integrate transcriptomes of cells, overcoming limitations of existing single-cell RNA sequencing technologies, and (2) relate cellular features to the histological and anatomical scales through integration in the *anatomical space*. We show how, starting from complementary experimental datasets obtained through different technologies characterized by different limitations, it is possible to address the individual technological limitations via integration. The proposed methodology is unsupervised and requires minimal domain knowledge.

5.1 Machine Learning-Driven Alignment of Spatially-Resolved Whole Transcriptomes with Tangram

We tackle integration challenges posed by the creation of high-resolution cell atlases from two complementary perspectives: (1) learning an alignment between experimental data measured through different technologies and (2) learning an anatomical manifold from a pre-existing atlas that allows the integration of the new data. For example, given a tissue for which we have collected single-cell RNA-sequencing data, Spatial Transcriptomics data (not necessarily at single-cell resolution) and histological images, we can (1) integrate single cells and Spatial Transcriptomics data, obtaining a high-quality cell-resolution map of all the genes for the tissue and (2) integrate the obtained transcriptomes to an existing organ-level or human-level cell atlas using the histological images. For both these directions the methodology proposed is fully data-driven, unsupervised, does not require previously defined rules and does not make strong assumptions about the available data sources. For example, the method is flexible with respect to the number and the type of measured genes, the available experimental sources, and the quality of the data (indeed, Tangram can improve data quality through integration).

Intuitively, the way Tangram achieves integration resembles a puzzle game. Tangram uses single cell RNA-sequencing data as "puzzle pieces" to align in space to match "the shape" of the spatial data. From the learned mapping function, Tangram can (1) expand from a measured subset of genes to genome wide profiles, (2) correct low-quality spatial measurements, (3) map and show the location of cells of different types in space, (4) infer the mixture of cells collectively measured through low resolution measurements, and (5) align multi-modal data at single cell resolution using transcriptomics data as a bridge. Technically, Tangram is based on the optimization through gradient descent of an "alignment" function that allows obtaining a "mapping", through which two complementary transcriptomics datasets are aligned. The integration at the histological and anatomical scales is achieved building a latent space as a proxy of a similarity metric through a Siamese neural network with a semantic segmentation algorithm.

Through a large set of experiments using various datasets measured with different technologies, we show how Tangram can be used to improve the resolution, the throughput, the quality and the available modalities of the starting datasets via integration. As a final case study we show how we can learn a histological and anatomical integrated atlas of the somatomotor area of the healthy adult mouse brain starting from publicly available datasets and atlases [1]. Additionally, Tangram has been used in conjunction with other computational methods to provide a global picture of the regulatory mechanisms that govern cellular diversification in the mammalian neocortex [2]. Details about the Tangram method and the experimental results can be found in [1] and in Chap. 6 of [12].

6 Conclusion

In this work we have investigated different facets of knowledge extraction from scientific data, focusing on ML methods and the specific challenges posed by scientific and experimental data analysis. Over the last few years, machine learning has been a game-changing technology across countless areas and fields. However, the specific features of scientific data demand to adapt these techniques to new requirements and roles. This research investigated several of these requirements, including the role of experimental uncertainty in data modeling, the ML-driven inference of biological systems properties, the role of ML in supporting data-driven scientific model development, and the ML-driven integration of complementary experimental technologies. The choice of considering multiple application scenarios (spanning chemistry, biology, and genomics) in this work derives from the objective of exploring these challenges under multiple point of views. Research on this topic is growing at a staggering rate and is opening up directions for many areas in ML.

References

1. T. Biancalani, G. Scalia, et al., *Deep Learning and Alignment of Spatially-Resolved Single Cell Transcriptomes with Tangram*. Accepted for publication in Nature Methods (2021)
2. D.J. Di Bella, E. Habibi, R.R. Stickels, G. Scalia, J. Brown, P. Yadollahpour, S.M. Yang, C. Abbate, T. Biancalani, E.Z. Macosko, F. Chen, A. Regev, P. Arlotta, Molecular logic of cellular diversification in the mouse cerebral cortex. Nature (2021)
3. E. Gawehn, J.A. Hiss, G. Schneider, Deep learning in drug discovery. Mol. Informatics **35**(1), 3–14 (2016)
4. I. Goodfellow, Y. Bengio, A. Courville, *Deep Learning* (MIT press, 2016)
5. C. Guo, G. Pleiss, Y. Sun, K.Q. Weinberger, On calibration of modern neural networks, in *Proceedings of the 34th International Conference on Machine Learning, ICML'17*, (2017), pp. 1321–1330
6. A. Kendall, Y. Gal, Y, What uncertainties do we need in Bayesian deep learning for computer vision?, in *Proceedings of the 31st International Conference on Neural Information Processing Systems, NIPS'17*, (2017), pp. 5580–5590
7. D. Lähnemann, J. Köster, E. Szczurek, D.J. McCarthy, S.C. Hicks, M.D. Robinson, C.A. Vallejos, K.R. Campbell, N. Beerenwinkel, A. Mahfouz et al., Eleven grand challenges in single-cell data science. Genome Biol. **21**(1), 1–35 (2020)
8. I.V. Pasquetto, B.M. Randles, C.L. Borgman, On the reuse of scientific data. Data Sci. J. **16**(8), 1–9 (2017)
9. B. Pernici, F. Ratti, G. Scalia, *About the Quality of Data and Services in Natural Sciences* (Springer International Publishing, Cham, 2021), pp. 236–248
10. F. Ratti, G. Scalia, B. Pernici, M. Magarini, A data-driven approach to optimize bounds on the capacity of the molecular channel, in *2020 IEEE Global Communications Conference (GLOBECOM)* (IEEE, 2020), pp. 1–7
11. A. Regev, S.A. Teichmann, E.S. Lander, I. Amit, C. Benoist, E. Birney, B. Bodenmiller, P. Campbell, P. Carninci, M. Clatworthy et al., Science forum: the human cell atlas. Elife **6**, e27041 (2017)
12. G. Scalia, Machine Learning-Driven Integration, Knowledge Extraction and Uncertainty Management for Scientific Data. Ph.D. thesis, Politecnico di Milano (2020)
13. G. Scalia, C.A. Grambow, B. Pernici, Y.P. Li, W.H. Green, Evaluating scalable uncertainty estimation methods for deep learning-based molecular property prediction. J. Chem. Information Modeling **60**(6), 2697–2717 (2020)
14. G. Scalia, M. Pelucchi, A. Stagni, A. Cuoci, T. Faravelli, B. Pernici, Towards a scientific data framework to support scientific model development. Data Sci. **2**(1–2), 245–273 (2019)
15. G. Scalia, M. Pelucchi, A. Stagni, T. Faravelli, B. Pernici, Storing combustion data experiments: new requirements emerging from a first prototype, in *Semantics, Analytics, Visualization, 3rd International Workshop, SAVE-SD 2017*, Revised Selected Papers, LNCS, vol. 10959 (Springer International Publishing, Cham, 2018), pp. 138–149
16. P. Schneider, W.P. Walters, A.T. Plowright, N. Sieroka, J. Listgarten, R.A. Goodnow, J. Fisher, J.M. Jansen, J.S. Duca, T.S. Rush, et al., Rethinking drug design in the artificial intelligence era, in *Nature Reviews Drug Discovery* (2019), pp. 1–12

17. T.J. Struble, et al., Current and future roles of artificial intelligence in medicinal chemistry synthesis. J. Med. Chem. (2020)
18. M.D. Wilkinson et al., The FAIR guiding principles for scientific data management and stewardship. Sci. Data **3**(160018), 1–9 (2016)

Telecommunications

Sensor-Assisted Cooperative Localization and Communication in Multi-agent Networks

Mattia Brambilla

Abstract This brief highlights research advances on cooperative techniques for localization and communication. These two macro trends are investigated in the general context of mobile multi-agent networks for situational awareness applications, where time-varying agents of unknown locations are asked to fulfill positioning and information sharing tasks. Cooperative localization is conceived for both active and passive agents, i.e., targets to be detected and localized, and it is analyzed in vehicular and maritime environments. Communication is investigated for vehicular scenarios, where vehicles are requested to share massive data in the perspective development of connected and automated mobility systems. Both research areas rely on the integration of heterogeneous sensors and communication. Specifically, it is studied how to improve localization by exploring communication techniques as well as how to enhance communication performances by extracting information from perception sensors. The dynamic environment of multi-agent systems calls for robust, flexible and adaptive techniques, capable of profitably fuse different types of information, and the outcomes of these researches show how a statistical approach based on cooperation guarantees higher resilience, reliability and confidence.

1 Introduction

This chapter summarizes the main research works characterizing my Ph.D. studies at Politecnico di Milano from Nov. 2017 until Oct. 2020, under the supervision of Professor Monica Nicoli, which lead to the doctoral dissertation in [1] and the scientific publications in [2–12].

In the context of multi-agent networks, this chapter discusses research advances on cooperative localization and communication among time-varying agents of unknown positions. These two tasks (localization and communication) may sound as independent and disjoint, however the digital evolution of communication technologies and

M. Brambilla (✉)
Dipartimento di Elettronica, Informazione e Bioingegneria, Politecnico di Milano, Via Ponzio 34/5, 20133 Milano, Italy
e-mail: mattia.brambilla@polimi.it

© The Author(s) 2022
L. Piroddi (ed.), *Special Topics in Information Technology*,
PoliMI SpringerBriefs, https://doi.org/10.1007/978-3-030-85918-3_11

hardware equipment enables their fruitful integration, leading them to mutually assist each other.

The intuition that communication and localization are going to converge can be found in the activities of standardization groups for 5G, with the introduction of specific positioning-related signals in the 3GPP[1] Release 16. The trend gets higher attention for next releases of the communication standard (i.e., 6G), where a shift towards high-frequency communications will allow to extract accurate information from the environment and provide precise situational awareness for the user. The use of signals at millimeter-wave (30–100 GHz), sub-terahertz (100–300 GHz), terahertz (300 GHz–10 THz) and optical (>10 THz) frequencies defines new frontiers for short-range wireless communications, opening poorly-explored areas over large scale and introducing new challenges for hardware, algorithms and protocol design.

Integration goes together with cooperation, as multiple connected devices (such as vehicles) engage into an evolving network characterized by intermittent links due to mobility. This scenario paves the way for cooperative processing algorithms where multiple types of information/data/measurements need to be constructively handled for a coherent analysis.

In the following, cooperative localization is addressed in Sect. 2, where a general formulation of the whole problem is given by specifically defining the modeling assumptions (Sect. 2.1) and the types of measurements (Sect. 2.2), followed by the conceptualization of the methodology (Sect. 2.3) and the definition of analyzed scenarios (Sect. 2.4). The topic of communication, instead, is discussed for the vehicular environment in Sect. 3, where beam-based communications are investigated. A same structure is used, starting from the main assumptions (Sect. 3.1), passing through the specification of vehicle measurements (Sect. 3.2), until the discussion on the methodology (Sect. 3.3) and simulation scenarios (Sect. 3.4). Extra research topics of my Ph.D. are referenced in Sect. 4. Concluding remarks are drawn in Sect. 5.

2 Cooperative Localization

The first major research topic of my Ph.D. studies covers the area of cooperative localization, and most of the contents have been published in [2–4]. A network of multiple agents has the advantage of overcoming the ego-agent capability by enlarging the set of available information. The performance of an individual localization system is limited both in terms of field of view and accuracy, as it can only rely on the information coming from a single device. On the other hand, connected agents can acquire more information from different perspectives, thus being able to provide a more accurate positioning. Localizing agents in a network refers to the capability of estimating their positions (and possibly other kinematics or non-kinematics parameters). This task involves both cooperative (which intentionally share information) and non-cooperative agents, the latter are called targets. The time-variant network

---

[1] Third Generation Partnership Project.

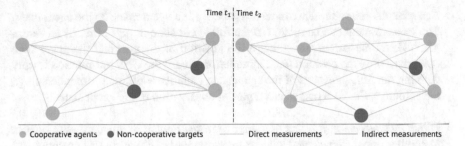

Time t_1 | Time t_2

● Cooperative agents ● Non-cooperative targets ——— Direct measurements ——— Indirect measurements

Fig. 1 Illustrative example of a time-varying topology of a generic multi-agent network of cooperative agents and non-cooperative targets. From time step t_1 to t_2, the mobility of agents and targets, as well as the availability of connections/measurements define two different network topologies

topology is created according to the agent/target dynamics (i.e., if they are static or moving) together with the availability/non-availability of connections among them. These concepts are graphically illustrated in Fig. 1, where an example of network topology variation between two time steps is provided.

2.1 Cooperative Localization: Assumptions

The research on cooperative localization is carried out under the following main conceptual assumptions (related to the system model). Agents and targets are assumed to be point objects in a 2D space, i.e., neglecting physical dimension and occupancy. It follows that the state of an agent comprises the position (and velocity) of the corresponding point. Practically, this choice can be referred to the use of the barycenter of the object itself. Ignoring the vertical dimension, instead, reduces the scene to be depicted over a plane, meaning that all agents and targets have a same height. All measurements are thus computed in a 2D domain with respect to the point representations, and they are assumed to be unique, i.e., it is not possible to have more than one data per each time step measuring a same quantity. It follows that an agent can have up to one measurement of each type (see next section) for itself and each nearby agent/target at each specific time step. As far as the algorithmic part is concerned, many other specific assumptions are considered but, to avoid to go into technicalities, they are omitted here. For details, please refer to [1, Ch. 3-4].

2.2 Cooperative Localization: Measurements

Cooperative localization of agents and targets is enabled by the use of measurements. In a general problem formulation, it is possible to classify measurements into three categories as follows:

- *Ego measurement*: it refers to a measurement of an agent made by the agent itself. The measurement can be related to the full agent state or part of it. As an example, a GPS[2] measurement estimates the position of the agent. If the agent state comprises only the position (e.g., latitude/longitude) the GPS information is truly defining the full agent state, if the agent state includes additional parameters (e.g., the velocity), the GPS data is just a partial observation of the agent state.

- *Direct measurement*: it refers to a relative measurement of an agent with respect to another one. Distance and angular measurements fall within this category. An agent is able to estimate the relative distance or bearing angle with respect to another one by decoding known signals. Specifically, considering a pair of agent where one plays the role of transmitter (Tx) and the other of receiver (Rx), the Rx-agent can retrieve its distance information with respect to the Tx by measuring the time difference between the signal generation and reception. Similarly, by estimating the direction of arrival of the received signal, the Rx-agent can determine the relative bearing with respect to the Tx one.

- *Indirect measurement*: it refers to a measurement of a target made by an agent. This type of measurements is used to localize non-cooperative targets that do not deliberately transmit a signal. Despite being similar to a direct measurement in the sense of letting an Rx-agent extracting range and bearing from a known signal, they significantly differ from the former as the measured quantities refer to an hypothesized target. This means that the Rx-agent has to decode a known signal sent from a Tx-agent, and discriminate between the direct path (i.e., the direct measurement) and the reflections. Identifying the source of reflections points let the Rx-agent determine the presence of unknown entities (i.e., the targets) in an indirect way. In this process, only first order reflections of the signal are used.

Note that a same known Tx signal can be exploited for both direct and indirect measurements, and the Rx capability of signal processing for multipath identification allows the distinction between direct and indirect measurement. Note also that Tx and Rx agents can coincide, meaning that the agent intentionally transmit a known signal for target detection and localization (in this specific case, a direct measurement does not exist). Technically, if Tx and Rx agents coincide, the configuration is monostatic, while if they differ it is bistatic.

2.3 Cooperative Localization: Methodology

The problem of cooperative localization of agents and targets by exploring the shared set of measurements described in Sect. 2.2 under the assumptions in Sect. 2.1 is addressed by developing a Bayesian filtering approach. The message passing scheme

---

[2] GPS—Global Positioning System.

of the sum-product algorithm (or belief propagation [13, 14]) is the algorithmic fundamental, in which measurements are converted into messages that bounce from agent-related, target-related and measurement-related variables in a loopy way and iteratively refine the agent/target beliefs. A belief is the estimation of agent/target marginal probability density distribution (e.g., the spatial area in which it is likely to be located). Message passing algorithms make use of a graphical representation of the multi-agent network to solve the considered inference problem, incorporating multiple variables into one unique framework. Graph-based models are suitable for capturing the intrinsic network-like structure of the multi-agent network, where agents and targets play the roles of nodes, while the connections among them indicate the availability of communication links and measurements. Such formulation has the advantage of fitting both centralized and distributed architectures. In the latter case, it allows for parallel information exchange (messages), speeding up the flooding of information (at each node) to rapidly reach all nodes in few multi-hop exchanges.

The proposed solution considers the localization of cooperative agents and the multi-target tracking as two integrated tasks. Compared to state-of-the-art solutions [15–19], the developed methodology integrates the uncertainties of agents (coming from the need of localizing them) with the tracking of unknown and arbitrary number of targets, and fruitfully exploits target localization as a mean to improve agent self-localization. The idea can be conceptualized as a dual layer mutually-connected graph (see Fig. 2), where the two tasks are indicated over two distinct planes but the upside-down connections among agents and targets denote a mutual exchange of information. Ego and direct measurements are used for the top layer (cooperative agent localization) only, while indirect measurements define the inter-layer connections and enable the multi-target tracking.

The proposed general formulation manages the agent/target uncertainties as well as communication failures (i.e., intermittent measurement availability) and typical multitarget tracking challenges like the target birth and death, the presence of clut-

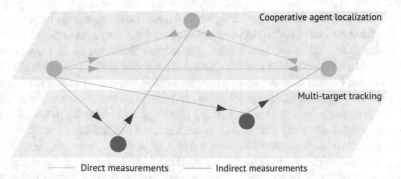

Fig. 2 Conceptualization of the integrated tasks of cooperative agent localization and multi-target tracking, which are unified in a single graph where messages (indicated as directed arrows) are flooded among agents and targets. The directions of the arrows from targets to agents indicate that target-related information is extracted and used to refine the beliefs of agents

ter measurements (i.e., false alarms), missed detections, and measurement origin uncertainty (i.e., the problem of unknown association between targets and indirect measurements). All details concerning the general stochastic modeling and algorithmic properties are detailed in the doctoral dissertation [1, Ch. 3].

2.4 Cooperative Localization: Application Scenarios

The developed solution for cooperative agent localization and multitarget tracking is general enough to accommodate a full multistatic network configuration, with multiple transmitters and receivers, thus guaranteeing a wide range of applications in safety-related and situational awareness contexts.

In my Ph.D. studies, researches have been focused on maritime and vehicular environments. In the former case (see [1, Ch. 3]), agents can be identified as autonomous underwater vehicles, ships/vessels, sea sensors, buoys or wave gliders, while targets can be intruding/illegal ships, military vessels or other types of potential threats. In the latter case (see [1, Ch. 4]), agents can be land vehicles, while targets can refer to vulnerable road users or other non-connected vehicles. Each environment has its own peculiarities, especially related to the agent/target mobility, technology, measurement availability and accuracy. Nevertheless, the proposed solution is flexible enough to accommodate for specific needs of each scenario, and it guarantees a high degree of adaptability.

The vehicular environment has been considered for the first analyses, and the achieved results considered a simplified algorithm for multitarget tracking without clutter and missed detections, with a known set of possible targets to be localized, and without direct measurements among vehicles. On the other hand, both centralized and distributed implementations have been tested, showing that a fully distributed vehicular network with direct communication links among vehicles can achieve a same performance as the case of a centralized architecture. In this scenario ego measurements are available through GPS, while target localization can be performed by processing on-board perception sensors (e.g., radar, lidar, camera), and the communications can exploit wireless links (e.g., cellular technology). The assessments through simulations consider both simplified scenes and realistic urban traffic flow conditions over a real road network, where two distinct mobility scenarios (a traffic light regulated case and a fully autonomous one) are taken into account. Performance results highlight the potential impact of the proposed technique for next generation mobility systems, showing enhancements with respect to GPS-based positioning.

The maritime environment has been studied as a second research area, where the generalized version of the proposed methodology has been truly applied for a centralized network architecture. This research comes from a collaboration with the NATO STO CMRE (North Atlantic Treaty Organization–Science and Technology Organisation–Centre for Maritime Research and Experimentation). In this scenario, ego measurements can be available through GPS in case of surface agents, while inertial information should be used in case of underwater agents. Due to water medium attenuation, in fact, it is only possible to use acoustic communication links to exchange information in the underwater domain. The same acoustic signal is used for direct and indirect measurements. The achieved results highlight the algorithm resilience to handle a time-varying number of mobile targets and extract from them a valuable information to improve the localization of agents, despite the challenging bistatic settings. It has been proven through simulations how the target implicit information is of utmost importance in case of agent outage conditions as it still allows to satisfactory localize *lost* agents. In the Ph.D. thesis, a simulated maritime surveillance use case is analyzed (see [1, Ch. 3]), but an assessment over real underwater data is going to be released too.

3 Vehicular Communication

The second major research topic of my Ph.D. studies covers the area of communication in next-generation vehicular networks, and most of the contents have been published in [5–10]. The need of letting vehicle share massive amount of data comes from the evolution towards connected and automated mobility systems, where a set of services are going to be available for users, improving road safety and transportation efficiency [20, 21]. Vehicle-to-Everything (V2X) communications require technologies, standards and protocols for the development of a connected mobility ecosystem. Vehicle-to-Vehicle (V2V) and Vehicle-to-Infrastructure (V2I) fall within the V2X umbrella. As the definitions may suggest, these specific types of wireless communications allow a vehicle to connect with nearby road players such as other vehicles (V2V), road infrastructure (V2I) or other types of actors like pedestrians (technically a device handled by a pedestrian, e.g., a smartphone). The research on vehicular communication is centered on beam-based communications, where vehicles use directive narrow beams to confine the signal with the assistance of on-board sensors (see Fig. 3). Two communication technologies are analyzed and compared, namely millimeter Wave (mmWave) and Free-Space Optics (FSO).

3.1 Vehicular Communication: Assumptions

This research addresses the currently emerging area of V2X communication, which is a rapidly-evolving field where brand new ideas are being proposed, also confirmed by an increasing interest from both research and industry perspectives, due to the needs of developing ad-hoc solutions to meet a constantly growing market demand. The addressed V2X-related research started from simplistic models of vehicles and trajectories, until reaching nearly realistic simulation settings. As a matter of facts, first analyses dealt with a 2D vehicle model [5, 6], as they were focused on preliminary assessment and validation of the impact of vertical vehicle vibrations (due to driving style and road pavement). Motivated by promising results and feedbacks, the

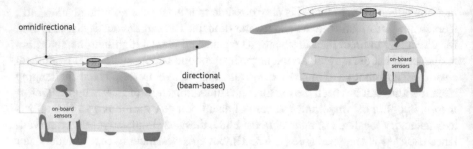

Fig. 3 Illustration of vehicular omnidirectional and beam-based communications. A connection with telecommunication apparatus and on-board sensors is also shown to highlight the concept of the proposed solution

complexity of system model has been increased by introducing a full 3D description of vehicle position and orientation [7, 8]. This generalization allowed more in-dept analyses on the degradation effects due to beam mis-alignment as a consequence of the combined effects of position and orientation errors, vehicle motion trajectory and road pavement imperfections. From a technological perspective, mmWave and FSO communication technology are simulated by assuming steerable devices able to irradiate in any 3D direction, and the V2V channels are always assumed as ideal (i.e., without obstructions, shadowing, fading, reflections, etc.). Indeed, the research focuses on the analysis of the degrading impact of misalignment on beam-based communications, thus separating the effects of signal transmission with the ones merely due to incorrect beam pointing.

3.2   Vehicular Communication: Measurements

The proposed sensor-assisted methodology for vehicular communications relies on the ego-vehicle knowledge of its position and orientation, which are used to control the beam. A vehicle uses on-board GPS and Inertial Measurements Unit (IMU) to retrieve its own position and orientation, to be possibly integrated in a navigation filter for a more robust estimation. The processed state information is locally used to adjust the beam pointing direction, i.e., by keeping the desired pointing regardless of the vehicle motion. Vehicle position information is also shared (over a reliable communication link) to enable cooperation with connected vehicles (which perform dual operations). Information sharing is thus the enabler of the cooperative sensor-assisted beam alignment scheme which is discussed in my doctoral dissertation [1, Ch. 5]. Therein, a system level architectural vision is also highlighted, offering a more general framing of the proposed solution in a realistic V2X connected mobility ecosystem. In particular, it is mentioned how it could be possible to integrate vehicle perception sensors (e.g., radar, lidar, camera) in an extension of the proposed solution.

3.3 Vehicular Communication: Methodology

Communicating at narrow beams is currently deemed as the most promising techno-logical solution able to meet high-data rate requirements for advanced autonomous automotive systems [22–25]. This research focuses on the use of on-board sensors (that precisely estimate the vehicle position and orientation) to assist the pointing of the beam. As communication involve two entities (a transmitter and a receiver), it is proposed to cooperatively share the vehicle state for a seamless and dynamic beam pointing, able to dynamically compensate for the reciprocal motion of vehicles. Being able to proactively known the position of the communication counterpart allows a vehicle to determine the candidate pointing direction without iteratively searching for the best candidate beam, thus saving time and improving the communication efficiency. Moreover, if position information comes with the associated accuracy, a vehicle can also implement a beamwidth and power adaptation algorithm, where not only the pointing direction is dynamically computed, but also the size of the beam and the required transmission power are optimally adjusted [10]. The exchange of vehicle navigation information is used as enabler to fasten the alignment of narrow beams such that the higher data rates of high frequency (mmWave and FSO) links are fully exploited.

Most of the research work is focused on the V2V link degradation due to beam misalignment as a consequence of vehicle mobility [5–8, 10], but a study on V2I channel modeling has also been carried out to evaluated potential benefits of exploit-ing mmWave sparsity to design a robust beam alignment [9].

3.4 Vehicular Communication: Application Scenarios

To evaluate the performance of the proposed sensor-assisted solution for beam-based vehicular communications, different simulation scenarios have been considered. As a starting point, a straight road use case has been the first considered environment for 2D evaluation [5, 6]. It has then been extended to account for full 3D mobil-ity over a winding trajectory [7, 8]. Analyses not included in the thesis but carried out in the last period of my Ph.D., instead, considered a set of collected mobility data [10]. In all cases, despite considering different environments, simulation con-ditions, technologies and assumptions, it has been shown that the proposed sensor-assisted beam-based (mmWave or FSO) solution for V2V communications brings remarkable improvements in terms of higher throughput, less outage and reduced signaling, providing an alternative to the onerous exhaustive beam search which does not use any information from sensors [1, Ch. 5].

4   Further Research Topics

During my Ph.D., I have also worked on extra topics that are not included in the doctoral dissertation. The interested reader in localization and tracking experimentation for industry application based on ultra wide band technology can find details in [11, 12].

5   Concluding Remarks

In this chapter, I summarized the major contents and outcomes of my doctoral research on sensor-assisted techniques for cooperative localization and communication. Multi-agent networks are the targeted use cases of this research, with main (but not exclusive) applications to maritime and vehicular contexts.

In the field of cooperative localization, graph theory has been used to describe the relations among agents, where the existence/absence of links allows/precludes the exchange of information, enabling a robust/weak data fusion for localization. Both centralized and distributed implementations have been considered, where the distinction reflects the specific targeted applications. Fully distributed solutions better suit vehicular networks, which are characterized by fast (order of seconds or less) variation in network topology, and where latency constraints call for decentralized processing at vehicles. On the other hand, in maritime surveillance, a different time scale (order of minutes) facilitates a centralized solution, where data from multiple agents/sensors are collected in a single fusion center. Moreover, processing offload is preferred due to intrinsic autonomy limitation of battery-powered vehicles, which are preferable to be long-lasting sensors rather than fast-processing units.

In the field of vehicular communication, it has been proposed to integrate on-board (already available) sensors to assist and enhance the communication performance. Guaranteeing a stable and reliable V2X link allows vehicles to implement advanced driving functionalities towards fully connected and automated mobility. It is stressed how the integration of two engineering domains (automotive and telecommunication) is highly recommended for the evolution of connected mobility services.

Recalling the introduction of this chapter, an interesting, pivotal and of utmost interest research direction would be the integration of the two fields in one joint localization and communication solution in the perspective of 6G wireless systems for integrated mobility over multiple domains, suitable for both centralized and distributed implementations. 6G wireless systems aim at the convergence of communication, sensing and localization in one integrated platform connecting multiple agents of diverse types, hardware and requirements, and the research topics in my Ph.D. thesis might inspire ideas for the standardization of new protocols and algorithms.

References

1. M. Brambilla, Sensor-assisted cooperative localization and communication in multi-agent networks. PhD dissertation, Politecnico di Milano (2021) doi: hdl.handle.net/10589/170700
2. M. Brambilla, G.M. Soatti, M. Nicoli, Precise vehicle positioning by cooperative feature association and tracking in vehicular networks, in *IEEE Statistical Signal Processing Workshop (SSP)* (2018), pp. 648–652. https://doi.org/10.1109/SSP.2018.8450794
3. M. Brambilla, M. Nicoli, G. Soatti, F. Deflorio, Augmenting vehicle localization by cooperative sensing of the driving environment: insight on data association in urban traffic scenarios. IEEE Trans. Intelligent Transp. Syst. **21**(4), 1646–1663 (2020). https://doi.org/10.1109/TITS.2019.2941435
4. R. Mendrzik, M. Brambilla, C. Allmann, M. Nicoli, W. Koch, G. Bauch, K. LePage, P. Braca, Joint multitarget tracking and dynamic network localization in the underwater domain, in *IEEE International Conference on Acoustics, Speech and Signal Processing (ICASSP)* (2020), pp. 4890–4894. https://doi.org/10.1109/ICASSP40776.2020.9054047
5. M. Brambilla, M. Nicoli, S. Savaresi, U. Spagnolini, Inertial sensor aided mmWave beam tracking to support cooperative autonomous driving, in *IEEE International Conference on Communications Workshops (ICC Workshops)* (2019), pp. 1–6. https://doi.org/10.1109/ICCW.2019.8756931
6. M. Brambilla, A. Matera, D. Tagliaferri, M. Nicoli, U. Spagnolini, RF-assisted free-space optics for 5G vehicle-to-vehicle communications, in *IEEE International Conference on Communications Workshops (ICC Workshops)* (2019), pp. 1–6. https://doi.org/10.1109/ICCW.2019.8757059
7. M. Brambilla, D. Tagliaferri, M. Nicoli, U. Spagnolini, Sensor and map-aided cooperative beam tracking for optical V2V communications, in *IEEE 91st Vehicular Technology Conference (VTC2020-Spring)* (2020), pp. 1–7. https://doi.org/10.1109/VTC2020-Spring48590.2020.9129590
8. M. Brambilla, L. Combi, A. Matera, D. Tagliaferri, M. Nicoli, U. Spagnolini, Sensor-aided V2X beam tracking for connected automated driving: distributed architecture and processing algorithms. Sensors **20**(12) (2020). https://doi.org/10.3390/s20123573
9. M. Brambilla, D. Pardo, M. Nicoli, Location-assisted subspace-based beam alignment in LOS/NLOS mm-wave V2X communications, in *IEEE International Conference on Communications (ICC)* (2020), pp. 1–6. https://doi.org/10.1109/ICC40277.2020.9148587
10. D. Tagliaferri, M. Brambilla, M. Nicoli, U. Spagnolini, Sensor-aided beamwidth and power control for next generation vehicular communications. IEEE Access **9**, 56301–56317 (2021). https://doi.org/10.1109/ACCESS.2021.3071726
11. L. Barbieri, M. Brambilla, R. Pitic, A. Trabattoni, S. Mervic, M. Nicoli, UWB real-time location systems for smart factory: augmentation methods and experiments, in *IEEE 31st Annual International Symposium on Personal, Indoor and Mobile Radio Communications* (2020), pp. 1–7. https://doi.org/10.1109/PIMRC48278.2020.9217307
12. L. Barbieri, R. Brambilla, A. Trabattoni, S. Mervic, M. Nicoli, UWB localization in a smart factory: augmentation methods and experimental assessment. IEEE Trans. Instrumentation Measurement (70) (2021) https://doi.org/10.1109/TIM.2021.3074403
13. F.R. Kschischang, B.J. Frey, H.A. Loeliger, Approximate evaluation of marginal association probabilities with belief propagation. IEEE Trans. Information Theor. **47**(2), 498–519 (2001). https://doi.org/10.1109/18.910572
14. J. Williams, R. Lau, Approximate evaluation of marginal association probabilities with belief propagation. IEEE Trans. Aerosp. Electronic Syst. **50**(4), 2942–2959 (2014). https://doi.org/10.1109/TAES.2014.120568
15. V. Savic, Hlawatsch, E.G. Larsson, Target tracking in confined environments with uncertain sensor positions. IEEE Trans. Vehicular Technol. **65**(2), 870–882 (2015). https://doi.org/10.1109/TVT.2015.2404132

16. F. Meyer, P. Braca, P. Willett, F. Hlawatsch, A scalable algorithm for tracking an unknown number of targets using multiple sensors. IEEE Trans. Signal Process. **65**(13), 3478–3493 (2017). https://doi.org/10.1109/TSP.2017.2688966

17. F. Meyer, T. Kropfreiter, J. Williams, R. Lau, F. Hlawatsch, P. Braca, M.Z. Win, Message passing algorithms for scalable multitarget tracking. Proc. IEEE **106**(2), 221–259 (2018). https://doi.org/10.1109/JPROC.2018.2789427

18. G. Soldi, F. Meyer, P. Braca, F. Hlawatsch, Self-tuning algorithms for multisensor-multitarget tracking using belief propagation. IEEE Trans. Signal Process. **67**(15), 3922–3937 (2019). https://doi.org/10.1109/TSP.2019.2916764

19. P. Sharma, A. Saucan, D.J. Bucci, P.K. Varshney, Decentralized Gaussian filters for cooperative self-localization and multi-target tracking. IEEE Trans. Signal Process. **67**(22), 5896–5911 (2019). https://doi.org/10.1109/TSP.2019.2946017

20. ETSI TR 102 638 v1.1.1: Intelligent Transport Systems (ITS); Vehicular Communications; Basic Set of Applications; Definitions (2009)

21. 3GPP TS 22.186 v16.2.0: 3rd Generation Partnership Project; technical specification group services and system aspects; study on enhancement of 3GPP support for 5G V2X services (Release 16) (2019)

22. B. Bertenyi, 5G evolution: what's next? IEEE Wireless Commun. **28**(1), 4–8 (2021). https://doi.org/10.1109/MWC.2021.9363048

23. H. Tataria, M. Shafi, A.F. Molisch, M. Dohler, H. Sjöland, F. Tufvesson, 6G wireless systems: vision, requirements, challenges, insights, and opportunities. Proc. IEEE **70**(7), 1166–1199 (2021). https://doi.org/10.1109/JPROC.2021.3061701

24. W. Jiang, B. Han, M.A. Habibi, H.D. Schotten, The road towards 6G: a comprehensive survey. IEEE Open J. Commun. Society **2**, 334–366 (2021). https://doi.org/10.1109/OJCOMS.2021.3057679

25. C. De Lima, D. Belot, R. Berkvens, A. Bourdoux, D. Dardari, M. Guillaud, M. Isomursu, E.S. Lohan, Y. Miao, A.N. Barreto, M.R.K. Aziz, J. Saloranta, T. Sanguanpuak, H. Sarieddeen, G. Seco-Granados, J. Suutala, T. Svensson, M. Valkama, B. Van Liempd, H. Wymeersch, Convergent communication, sensing and localization in 6G systems: an overview of technologies, opportunities and challenges. IEEE Access **9**, 26902–26925 (2021). https://doi.org/10.1109/ACCESS.2021.3053486

Design and Control Recipes for Complex Photonic Integrated Circuits

Maziyar Milanizadeh

1 Introduction to Photonic Integrated Circuits

In the last decades optical communications contributed to the huge diffusion of the telecommunication market, pushing the development of new technologies to enable higher performances and lower costs. We are witnessing the introduction of the light at every level of the communication network. Long distance links, national backbones and metropolitan networks are equipped with optical fibers that guarantees higher performances and lower costs than older technologies, e.g. radio or coaxial links. In Europe, U.S. and Japan there are projects to introduce fibers also in the short distance links (Next Generetion Access Network) ensuring a great rise of the bandwidth for the final consumer user. Optical bus will be introduced in the next future in the rack backplanes, on the boards and finally also into the chips on the boards, joining electronic devices and substituting them in the information delivering.

Then the photonic technologies are in a very predominant position in the today telecommunication scenario, that becomes the driving force behind the development of photonic integration. For example, optical components become the enabling factors for the optical networks when a complexity enhancement is required (e.g. where there is the necessity to move from point-to-point links to more complex network structures, introducing advanced all-optical switching technologies). Obviously not only telecommunication area takes advantage of this progress in technology but also different fields such as sensor for physical, medical or civil monitoring.

Photonic integration has a very long history dating back to the '70s. Starting from the first products the trend continued with more complicated structures bringing, in the present day, to the possibility to implement complex functions in very small chips, following more or less the path traced by the electronics some decades before. But at the moment the research on optical components is late compared with the market

M. Milanizadeh (✉)
Dipartimento di Elettronica, Informazione e Bioingegneria, Politecnico di Milano, Via Ponzio 34/5, 20133 Milano, Italy
e-mail: maziyar.milanizadeh@polimi.it

© The Author(s) 2022
L. Piroddi (ed.), *Special Topics in Information Technology*,
PoliMI SpringerBriefs, https://doi.org/10.1007/978-3-030-85918-3_12

requests, which show the necessity to integrate a lot of functions on a single chip with very low costs. This generate the great push toward the photonic integration, in particular for PIC with high complexity (in terms of number of components per PIC).

Nowadays, integrated photonics technologies are envisioned as fundamental for applications such as optical communications [1–3], optical interconnects [4–6], bio-sensing [7–9], 5G networks [10, 11] and quantum photonics [12].

Among all the technological platforms that can be used to realize PIC, two semi-conductors technologies have been emerging in the last decade: Indium Phosphide and Silicon. Thanks to the high index contrast offered by these photonic platforms, it is possible to integrated a large number of devices on the same chip and to implement complex functionalities for the generation, manipulation and detection of light. Indium phosphide offers the possibility of monolithically integrating on-chip waveguide, detectors, modulators and light sources [13], while silicon allows an unprecedented number of integrated components [14].

To reach the objective of replacing electronic circuits with low power consumption photonic circuits, especially for telecom/datacom applications [15, 16], it is necessary to handle more and more complex functions in the optical domain [17]. To realize such complex functions, photonic circuits must reach an higher level of complexity, interconnecting many photonic devices on the same chip.

Despite the device miniaturization achievable with silicon photonics technologies [18], the integration of these devices is a separated issue [19] that must be challenged to deliver advanced functionalities on photonic integrated chips [18]. This scaled complexity must be matched with the urgent needs of adaptability and programmability to enable the realization of arbitrary, reconfigurable, complex circuits thus shifting the paradigm from a device-level to a "system-on-a-chip" one.

1.1 The Control Paradigm

The extreme device miniaturization reached by state-of-the art photonic technologies now enables the realization of hundreds or even thousands of photonic elements in a footprint of less than 1 mm^2 [14]. Although many building blocks potentially provide the required degrees of freedom to realize flexible and arbitrarily complex photonic architectures, reconfigurable optical circuits aggregating many different functionalities are still encountering strong difficulties to emerge. The reason is that in photonics, similarly to electronics, device miniaturization is not synonymous with large scale of integration, and some keys still need to be found to make photonics step up from the current device level to complex, adaptive and reconfigurable integrated circuits.

In other side, PICs are evolving towards on-chip re-configurable architectures and general purpose programmable photonic processors, enabling the implementation of many different functionalities on-demand [20–23]. These schemes rely on the use of a large number of optical interferometers, such as MZI and MRRs, whose individual working point is inherently related to the phase delay between the interfering optical beams. Therefore, any kind of phase perturbation may substantially affect the overall behavior of the PIC.

To reach these goals, feedback control is mandatory to steer and hold the entire system to the desired functionality, and make it immune to fabrication tolerances, functional and environmental drifts, and mutual crosstalk effects. In fact sensitivity to temperature fluctuations is one of the strongest limiting factors to the exploitation of integrated optical devices. This effect is particularly relevant in Silicon on Insulator (SOI), where the large thermo-optic coefficient (TOC) of silicon ($1.8 * 10^{-4} K^{-1}$ at 300 K [25]) is responsible for a wavelength shift of any interference-based device of about $10 GHz \cdot K^{-1}$. In interferometric devices, a waveguide width deviation of only 1 nm can produce a frequency shift of about 100 GHz in the spectral response [24]. Due to this sensitivity, and to the tolerances of current fabrication technologies, the response of fabricated PIC hardly matches the design performance; moreover, it is possible that reconfiguration capabilities will be needed to adapt in real time the circuit to new requirements (i.e. dynamic switching and routing, channels add/drop and so on). Due to functional drifts and components ageing over time, is not possible to have an accurate and robust control only relying on lookup tables; hence, there is a need of an automatic feedback loop to set the new working point automatically and to control real time the PIC to counteracts unwanted drifts.

The paradigm of a feedback loop for integrated photonics is shown in Fig. 1. One or more detectors are placed in strategic positions throughout the circuit; these detectors, which can be fast or slow depending on what is being measured, generate control signals proportional to the measure they are performing. A controller reads the control signals and estimates the working point of the PIC based on the information provided; the controller then drives the actuators to steer the working point accordingly to the algorithm implemented in the controller itself. The actuators rely on physical effects to modify the working point of the PIC. Conveniently, control systems should be low cost, energy efficient, insensitive to fluctuations of the optical power, applicable to both passive and active devices, and should not require additional photonic structures.

Optical Design and control approaches should bind together for development of complex reconfigurable PIC. This means along the design process we need to take

Fig. 1 Closed loop control of PIC. The signals from detectors placed in strategic positions throughout the circuits is used by a controller to estimate the current working point. Actuators are then driven to modify the working point of the circuit accordingly to the logic implemented in the controller

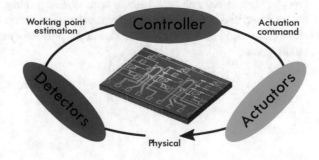

into account tunability of the device in different scenario of application and for variety of possible perturbations. In this work we study some of these techniques and introduce new approaches to design and control complex PICs, reconfigurable along the operation band.

2   Key Proposed Innovative Concepts

The work reported in this thesis has outcomes in diverse areas under the Integrated Photonics umbrella. Solutions and techniques from design to calibration and control phase of PIC implementation are discussed and analyzed through numerical simulations and experimental trials of different functions implemented in multiple technologies.

- **Thermal crosstalk free system**

We discuss extensively the idea of cross talk between effect of actuators in PIC. For the example of thermo-optic actuators, due to heat flow from actuator to other parts of chip, perturbation is introduced in unwanted places of circuit from each actuator. To mitigate this perturbations other heaters need to change their working point which lead to other perturbations which in some cases can cause instability in control algorithms. Several solutions have been proposed to mitigate thermal cross-talk on a photonic chip. Thermal isolation trenches allow the localization of the heat around the actuated waveguide and can also improve the heater efficiency [21, 26]. However, this approach puts some constraints on the layout and footprint of the PIC. In this work we suggest instead of individual modifications of actuators, they are modified by specific weights which are obtained from eigen-solution of the coupled system (introduced as Thermal eigenmode decomposition, TED). Through numerical simulations and experimental trials we examine the efficiency of this technique. We suggest an effective technique to evaluate coupled system to deliver the optimum cross-talk cancellation with TED based techniques.

A key point of the TED method is the knowledge of the **T** matrix describing the thermal coupled system. In principle it could be inferred from thermal simulations. However, this approach is hardly practical for large scale PICs accounting for many heat sources, especially because the overall assembly of the PIC should be considered, including on-chip metal lines, wire bonding, chip submount and ultimately the package itself. Therefore, for an accurate estimation of the **T** matrix, alternative strategies should be adopted, which should be preferably applicable to arbitrary PIC architectures.

We considered three different solutions to evaluate the thermal crosstalk with different accuracy:

1. Heuristic (lowest accuracy): an estimation of thermal cross talk is inferred by simply considering the topology of the circuit;
2. Optical measurement (highest accuracy, not always practicable): thermal crosstalk is directly estimated by measuring the unwanted wavelength shift of neighbor devices;
3. Electrical measurement (good accuracy, always practicable): thermal crosstalk is indirectly estimated by measuring the change of the electric resistance of neighbor heaters used as temperature probes;

It is worth to mention that T depends only on the device architecture and not on the specific operation and, as any transfer function, must be evaluated once, independently on the circuit working point.

A direct optical measurement of thermal crosstalk is not possible in most cases, because it requires the possibility to optically access the output ports of each tunable element. A good estimation of the T matrix can be achieved electrically, by measuring the temperature-induced change of the resistance of metallic probes located in suitable spots of the PICs. To this aim, we can advantageously use the heater structures themselves, which can be used also as temperature sensors to map thermal crosstalk across the photonic chip. We managed to demonstrate through experimental trials the effectiveness of this technique. In Fig. 2 we compare of convergence efficiency of TED based control algorithms with different techniques of estimation for T matrix for a cross-bar connect of 4 MRRs.

Details of these techniques can be found in published work [27, 29].

- **Tuning to signal spectrum**

  After introducing an effective techcnique to mitigate thermal cross-talk effects in PIC, we continue by discussing a new approach in tuning algorithm to identify the progress of the algorithm. Traditionally, PIC tuning is implemented targeting a specific frequency response. This approach can be valuable for testing, and pre-calibration procedures, but it is not practical for automated tuning of the PIC during its operation because it is resource and time consuming. For instance, it may require a tunable source and/or a spectrum analyzer to monitor the response at wavelengths of interest. Likewise, time domain approaches, based on the measurement of the Bit error rate (BER) or eye diagram distortion, require considerable load of Digital Signal Processor (DSP) as well as electrical power consumption. In any case both approaches are hardly practical when the device is in use. In this work we suggest using the Power Spectrum Density (PSD) of the channel as merit in tuning algorithm and we demonstrate through experimental trials we can tailor the frequency response of the filter, even if its not design to be modified, to the PSD of the channel, finding the best filter tuning condition for that specific channel. Using the optical power at the output of a golden filter (filter with acceptable frequency behavior) we can tune the same family of devices. In other words replication of filter frequency response which can save time and resource in calibration phase of PIC implementation. Through numerical simulations and experimental trials we demonstrated the effectiveness of this technique in tuning and tailoring PIC to different channels. Eventually, we discuss extension of this idea forcing devices to operate on specific channel through Wavelength Division Multiplexing (WDM) grid via labeling technique. Details can be found in [28, 29].

- **Automatic calibration and dynamic look up tables**
  Based on the concept of using signal optical power in tuning and filter replication, we introduce dynamic Look Up Tables (LUT) which can be automatically created for arbitrary PIC. They can be updated due to new requirements of the operation or new conditions of the chip like perturbations from neighboring circuits. Traditionally PICs are kept in their optimum working point through pre-defined tables which are expensive to produce during the calibration of the device. These tables need to be updated due to aging of the device or new requirement of the operations. We suggest adopting TED-based algorithms while using optical power of the channel to automatically create these tables and through locking algorithms these LUTs are updated matching new conditions of operations.

- **Reduction of electrical I/O**
  Electrical bonding pads, responsible for electrical I/O occupy huge area of photonic chips, their size can not be decreased due to mechanical limits of wire bonding techniques [30]. This dictates complex assembly techniques for large PICs like flip chip [31] approaches which other than being expensive have their limitations. In this work, we suggest two approaches to reduce the electrical connection needed to operate and control PICs. We discuss implementing electronic multiplexer in optical chip to reduce needed electrical lanes as the first solution and grouping together optical waveguides and use common detector as the second one. In this approach the sum of optical power from monitor ports of PIC is measured and needed to be carried out of the chip. Labeled signals can be used to distinguish through a demodulation of the measured signal at the controller side out of the photonic chip.

- **FSR free and hitless-tunable Polarization-Diversity filter**
  We introduce a fully-reconfigurable add-drop filter based on high order coupled MRR with exceptional features suitable for dynamic bandwidth allocation in core networks, back-haul networks, intra- and inter-datacenter interconnects. Implemented in Silicon on Insulator platform, we recorded spurious free Free Spectrum

Range (FSR) on 100nm wavelength range centered in C band in both Drop and Through port. This is achieved using non-integer ratio Vernier scheme implemented through optimization techniques to design appropriate radius of MRR and their couplers. This device can be reconfigured transparently (hitless tuning) on this wide operation band in nano-second time scale without introduction of any perturbation (>35 dB channel isolation). This feature is obtained by implementing Variable Optical Attenuators (VOA) in middle MRR via P-i-N junctions. This filter is implemented in polarization diversity scheme (1.2 dB polarization dependent loss), demonstrating polarization transparency for the hole operation band. Its performance is evaluated in experimental trials specifically while adding and dropping two double polarization channels (100 Gbit/s (QPSK) and 200 Gbit/s (16-QAM)) with complex modulation scheme demonstrated in Fig. 3. Multiples of this filter can be implemented to operate on the same optical bus to add and drop channels from Dense Wavelength Division Multiplexing (DWDM) grid with operation range wider than extended C band. More information about this work can be found in [32].

- **Free space beam manipulation**
  This work is concluded by discussing the integrated meshes based on MZI units. We demonstrate manipulation of free-space optical beams by using a 4 × 1 self-configuring Silicon photonic mesh. In particular, after comparing variety of integrated mesh structures a self-converging control loop is suggested and examined capable of tuning and re-steering such devices. The performance of control approach is examined for the case of global detector (single PD) or spread detection (detector at each stage). Effect of these control algorithms on the radiation performance of the device are analyzed to understand the silver lining between sensitivity and perturbation. Then using a simple 4 × 1 diagonal mesh, we successfully performed beam coupling from a light source in the free space to a SMF just accurately tuning the MZIs mesh. As expected, also the reversed process, from a SMF to a specific position, worked as well. Steering the beam through mesh tuning is therefore a straightforward consequence of the alienable beam coupler when used in the opposite propagation direction. Capability of mesh to reconstruct the beam front was putted to trials by introducing known and unknown (replicating unwanted free space perturbations) perturbing mediums in the image and far filed plane of the device. In all the trials (within the degrees of freedom offered by this mesh) ideal beam pattern was recovered which demonstrated the possibility of mesh to reconstruct the optimum free space channel. Integrated meshes can be automatically tuned (by simply minimizing stages) to manipulate both phase and amplitude of the array elements to offer more degrees of modification in beam front comparing with phase array structures.

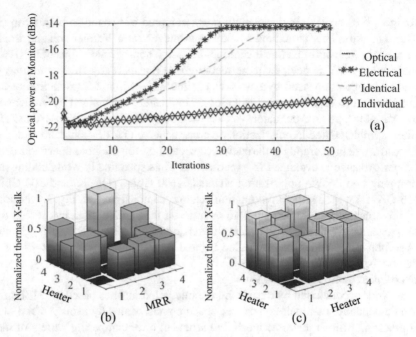

Fig. 2 **a** Convergence rate of TED-based tuning of the 2 × 2 MRR cross-bar interconnect when the off-diagonal terms of the phase coupling matrix T are assumed to be identical (green dashed), when they are estimated from electrical measurements (red asterisks), from optical measurements (blue solid). Black diamonds show the converge rate when MRRs are individually tuned (no-TED). **b** Optically and **c** electrically measured off-diagonal terms of the phase coupling matrix T normalized to the maximum value

Comparing with phase array antennas with these meshes not only an straight forward power minimization automatic tuning is achievable, due to the access to array elements amplitude, these circuits can obtain more sophisticated free space beams and introduce vast level of correction. The performance of the mesh in steering, coupling and identifying free-space beams can be improved by optimizing the design of the radiating elements, whose number can be scaled up without impairing the progressive self-configuration procedure employed for the tuning of the mesh. Applications are envisioned to more advanced free-space optical processing, including phase front reconstruction, beaming through scattering media and chip-to-chip free space communications. Details of these works can be found in [33–35].

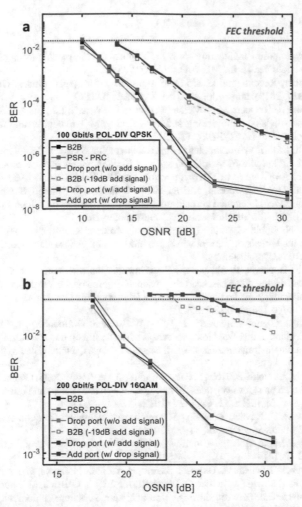

Fig. 3 BER versus Optical Signal to Noise Ratio (OSNR) of **a** 100 Gbit/s Pol-Div QPSK channel, **b** 200 Gbit/s Pol-Div 16QAM channel. Black curve for back to back scenario replacing the device with a direct fiber in the setup, Blue curve for connected device while the filters were de-tuned from the channel to measure the transparency of Pol-Div scheme, Yellow curve for dropped channel. Red and Pink curves are for adding and dropping of two channels at the same wavelength with the same modulation scheme being launched at In and ADD port. Dashed-Black curve is the performance of the transceiver for B2B condition when two channels are coupled by a fiber coupler and one (the perturbing channel) is 19 dB below the measuring one

References

1. C.R. Doerr, Proposed architecture for MIMO optical demultiplexing using photonic integration. IEEE Photonics Technol. Lett. **23**, 1573–1575 (2011)
2. D.A.B. Miller, Reconfigurable add-drop multiplexer for spatial modes. Opt. Express **21**, 20220–20229 (2013). https://doi.org/10.1364/OE.21.020220
3. N.K. Fontaine, C.R. Doerr, M.A. Mestre, R.R. Ryf, P.J. Winzer, L.L. Buhl, Y. Sun, X. Jiang, R. Lingle, Space-division multiplexing and all-optical MIMO demultiplexing using a photonic integrated circuit, in *OFC/NFOEC* (2012)
4. D.A.B. Miller, Device requirements for optical interconnects to CMOS silicon chips. PMB.3, in *Integrated Photonics Research, Silicon and Nanophotonics and Photonics in Switching, Optical Society of America* (2010). https://doi.org/10.1364/PS.2010.PMB3
5. P. Chaisakul, D. Marris-Morini, J. Frigerio, D. Chrastina, S. Rouifed, S. Cecchi, P. Crozat, G. Isella, L. Vivien, Integrated germanium optical interconnects on silicon substrates. Nature Photonics **8**, 482–488 https://doi.org/10.1038/nphoton.2014.73
6. L.-W. Luo, N. Ophir, C. Chen, L. Gabrielli, C. Poitras, K. Bergmen, M. Lipson, WDM-compatible mode-division multiplexing on a silicon chip. Nature Commun. **5** (2014) https://doi.org/10.1038/ncomms4069
7. T. Claes, W. Bogaerts, P. Bienstman, Experimental characterization of a silicon photonic biosensor consisting of two cascaded ring resonators based on the Vernier-effect and introduction of a curve fitting method for an improved detection limit. Opt. Express **18**, 22747–22761. https://doi.org/10.1364/OE.18.022747
8. M. Iqbal, M. A. Gleeson, B. Spaugh, F. Tybor, W.G. Gunn, M. Hochberg, T. Baehr-Jones, R.C. Bailey, L.C. Gunn, Label-free biosensor arrays based on silicon ring resonators and high-speed optical scanning instrumentation. IEEE J. Selected Topics Quantum Electronics **16**, 654–661 (2010)
9. V. Passaro, M. La Notte, B. Troia, L. Passaquindici, G. De Francesco, G. Giannoccaro, Photonic structures based on slot waveguides for nanosensors: state of the art and future developments. Int. J. Res. Rev. Appl. Sci. **11**, 411–427 (2012)
10. R. Waterhouse, D. Novack, Realizing 5G: microwave photonics for 5G mobile wireless systems. IEEE Microwave Mag. **16**, 84–92 (2015)
11. J. Capmany, P. Muñoz, Integrated microwave photonics for radio access networks. J. Lightwave Technol. **32** (2014)
12. J. Silverstone, D. Bonneau, K. Ohira, N. Suzuki, H. Yoshida, N. Iizuka, M. Ezaki, C. Natarajan, M. Tanner, R. Hadfield, V. Zwiller, G. Marshall, J. Rarity, J. O'Brien, M. Thompson, On-chip quantum interference between silicon photon-pair sources. Nature Photonics **8** (2014). https://doi.org/10.1038/nphoton.2013.339
13. M. Smit, et al., An introduction to InP-based generic integration technology. Semiconductor Sci. Technol. **29**, 083001 (2014)
14. J. Sun, E. Timurdogan, A. Yaacobi, E. Hosseini, M.R. Watts, Large-scale nanophotonic phased array. Nature (2013). https://doi.org/10.1038/nature11727
15. A. Vahdat, H. Liu, X. Zhao, C. Johnson, *The Emerging Optical Data Center.* (2011) https://doi.org/10.1364/OFC.2011.OTuH2
16. K. Bergman, J. Shalf, T. Hausken, Optical interconnects and extreme computing. Opt. Photon. News **27** (2016). https://doi.org/10.1364/OPN.27.4.000032
17. D.A.B. Miller, Device requirements for optical interconnects to silicon chips. Proc. IEEE **97**, 1166–1185 (2009)
18. T. Baehr-Jones, T. Pinguet, P. Guo-Qiang, S. Danziger, D. Prather, M. Hochberg, Myths and rumours of silicon photonics. Nature Photonics **6**, 206–208 (2012). https://doi.org/10.1038/nphoton.2012.66
19. R. Chau, B. Doyle, S. Datta, J. Kavalieros, K. Zhang, Integrated nanoelectronics for the future. Nature Materials **6** (2007). https://doi.org/10.1038/nmat2014

20. Y. Shen, N.C. Harris, S. Skirlo, M. Prabhu, T. Baehr-Jones, M. Hochberg, X. Sun, S. Zhao,H. Larochelle, D. Englund, et al., Deep learning with coherent nanophotonic circuits. Nature Photonics **11**, 441–446 (2017). https://doi.org/10.1038/nphoton.2017.93

21. D. Pérez, I. Gasulla, L. Crudgington, D.J. Thomson, A.Z. Khokhar, K. Li, W. Cao, G.Z. Mashanovich, J. Capmany, Multipurpose silicon photonics signal processor core. Nature Commun. **8** (2017). https://doi.org/10.1038/s41467-017-01529-w

22. J. Carolan, C. Harrold, C. Sparrow, E. Martín-López, N.J. Russell, J.W. Silverstone, P.J. Shadbolt, N. Matsuda, M. Oguma, M. Itoh, G.D. Marshall, M.G. Thompson, J.C.F. Matthews, T. Hashimoto, J.L. O'Brien, A. Laing, Anthony, Universal linear optics. Am. Assoc. Adv. Sci. (2015) https://doi.org/10.1126/science.aab3642

23. L. Zhuang C.G.H. Roeloffzen, M. Hoekman, K.-J. Boller, A.J. Lowery, Programmable photonic signal processor chip for radiofrequency applications. Optica **2**, 854–859 (2015). https://doi.org/10.1364/OPTICA.2.000854

24. F. Xia, L. Sekaric, Y. Vlasov, Ultracompact optical buffers on a silicon chip. Nature Photonics. https://doi.org/10.1038/nphoton.2006.42

25. J. Komma, C. Schwarz, G. Hofmann, D. Heinert, R. Nawrodt, Thermo-optic coefficient of silicon at 1550 nm and cryogenic temperatures. Appl. Phys. Lett. **101** (2012). https://doi.org/10.1063/1.4738989

26. P. Dong, W. Qian, H. Liang, R. Shafiiha, F. Ning-Ning, F. Dazeng, X. Zheng, A.V. Krishnamoorthy M. Asghari, Low power and compact reconfigurable multiplexing devices based on silicon microring resonators. Opt. Express **18**, 9852–9858 (2010). https://doi.org/10.1364/OE.18.009852

27. M. Milanizadeh, D. Aguiar, A. Melloni, F. Morichetti, Canceling thermal cross-talk effects in photonic integrated circuits. J. Lightwave Technol. **37**, 1325–1332 (2019)

28. D. Aguiar, M. Milanizadeh, E. Guglielmi, F. Zanetto, G. Ferrari, M. Sampietro, F. Morichetti, A. Melloni, Automatic tuning of silicon photonics microring filter array for hitless reconfigurable add-drop. J. Lightwave Technol. **37**, 3939–3947 (2019)

29. M. Milanizadeh, S. Ahmadi, M. Petrini, D. Aguiar, R. Mazzanti, F. Zanetto, E. Guglielmi, M. Sampietro, F. Morichetti, A. Melloni, Control and calibration recipes for photonic integrated circuits. IEEE J. Selected Topics Quantum Electronics **26**, 1–10 (2020)

30. K. Yu, C. Chen, C. Li, H. Li, A. Titriku, B. Wang, A. Shafik, Z. Wang, M. Fiorentino, P. Yin Chiang, S. Palermo, 25Gb/s hybrid-integrated silicon photonic receiver with microring wavelength stabilization, in *2015 Optical Fiber Communications Conference and Exhibition (OFC)*

31. X. Zheng, E. Chang, P. Amberg, I. Shubin, J. Lexau, F. Liu, H. Thacker, S.S. Djordjevic, S. Lin, Y. Luo, J. Yao, J.-H. Lee, K. Raj, R. Ho, J.E. Cunningham, A.V. Krishnamoorthy, A high-speed, tunable silicon photonic ring modulator integrated with ultra-efficient active wavelength control. Opt. Express **22**, 12628–12633 (2014). https://doi.org/10.1364/OE.22.012628

32. F. Morichetti, M. Milanizadeh, M. Petrini, F. Zanetto, G. Ferrari, D.O. de Aguiar, E. Guglielmi, M. Sampietro, A. Melloni, Polarization-transparent silicon photonic add-drop multiplexer with wideband hitless tuneability. Nature Commun. https://doi.org/10.21203/rs.3.rs-147585/v1

33. M. Milanizadeh, P. Borga, F. Morichetti, D.A.B. Miller, A. Melloni, Manipulating free-space optical beams with a Silicon photonic mesh, in *2019 IEEE Photonics Society Summer Topical Meeting Series (SUM)*

34. M. Milanizadeh, E. Damiani, T. Jonuzi, M.J. Mencagli, B. Edwards, D.A.B. Miller, N. Engheta, A. Melloni, F. Morichetti, Recursive MZI mesh for integral equation implementation, in *European Conference on Integrated Optics 2020 (ECIO)*

35. M. Milanizadeh, et al, Coherent self-control of free space optical beams with integrated silicon photonic meshes, in *Photonics Research*, under review (arXiv:2104.08174)

Printed in the United States
by Baker & Taylor Publisher Services